31865002779519
616.398 WAR
The truth about fat

WITHDRAWN

 River Forest Public Library
735 Lathrop Avenue
River Forest, IL 60305
708-366-5205
February 2019

THE TRUTH ABOUT FAT

Also by Anthony Warner

The Angry Chef:
Bad Science and the Truth About Healthy Eating

THE TRUTH ABOUT FAT

ANTHONY WARNER

ONEWORLD

A Oneworld Book

First published by Oneworld Publications Ltd, 2019

Copyright © Anthony Warner 2019

The moral right of Anthony Warner to be identified as the Author of this work has been asserted by him in accordance with the Copyright, Designs, and Patents Act 1988

All rights reserved

Copyright under Berne Convention

A CIP record for this title is available from the British Library

Hardback ISBN 978-1-78607-513-0
Trade Paperback ISBN 978-1-78607-593-2
eISBN 978-1-78607-514-7

Typeset by Palimpsest Book Production Limited, Falkirk, Stirlingshire
Printed and bound in Great Britain by Clays Ltd, Elcograf S.p.A.

Oneworld Publications Ltd
10 Bloomsbury Street
London WC1B 3SR
England

Stay up to date with the latest books,
special offers, and exclusive content from
Oneworld with our newsletter

Sign up on our website
oneworld-publications.com

MIX
Paper from
responsible sources
FSC® C018072

Doubt is not a pleasant condition, but certainty is absurd.

Voltaire

CONTENTS

Part III – What Should We Do?

A WORD ON LANGUAGE

Throughout this book I am going to talk about fat, obesity and body weight. In researching, I spent a great deal of time thinking about the language I was going to use, keen as I am not to add to the stigma that surrounds these issues. I appreciate the power of words, and I want to get it right.

After much consideration, I have decided to talk about 'fat people', 'people getting fat', and fatness generally, despite those terms being loaded with the potential to cause offence. I appreciate that 'fat' is often derogatory, but it remains better than the alternatives. I shall also refer to obesity and obese people, even though I am not completely happy with those terms, as we will discuss.

There are several reasons for these choices. Firstly, most of the alternatives are clunky and difficult to include consistently in prose. 'People in larger bodies', 'people with obese bodies', 'people suffering from obesity', or similar, are awkward and, by tiptoeing around the definition, draw attention to themselves. I believe that this would increase any stigmatising effect, and runs the risk of making me sound like a tool.

Also, although some people do prefer 'people first' language, I know many sensible commenters who do not, and I share the concerns they have with this approach. It seems to me that 'people first' language is only used if we consider the condition being described as undesirable, and if we want to decrease stigma, perhaps we should stop seeing fat in this way. Many autistic campaigners have long rallied against the use of 'people with autism', which makes it sound like a contagious disease, and I wonder if a similar approach might help when it comes to fat.

In truth, I have never met anyone offended by the use of the term 'fat people', but if you are, I can only apologise. I realise it is a difficult term for some, and may carry a painful history with it, but I was not aware of a practical alternative. It is common for stigmatised groups to reclaim insults so that they lose their power to hurt – it hasn't happened to 'fat' yet, but maybe now is a good time to try.

INTRODUCTION

Everyone knows we are in the grip of an obesity epidemic. It is estimated that by 2025, 18 per cent of the world's men and 21 per cent of the women will be obese. In the UK, in 2017 we were declared the fattest country in Western Europe, with 63 per cent of adults overweight, and 27 per cent of them obese.[1] And as obesity is strongly linked to a range of chronic diseases, its inexorable rise over the past forty years is likely to have a profound human cost.

You do not have to read the statistics to see that there is a problem. It is clear when you walk down the street in the developed world that people are getting larger, and have been for many years. It is a modern health crisis, unstoppably spreading around the world. As economies develop and Westernise, obesity sweeps in, making huge numbers of people fat and sick. To make matters worse, obesity is no longer just a disease of affluence. It is increasingly associated with poverty, inequality and divided societies.

According to everyone whose dietary beliefs I criticise, it is the fault of people like me. In supporting the consensus view of nutrition science, and particularly in my refusal to utterly condemn all convenience foods as the cause of this crisis, I am making the world fat. I have even publicly stated that sugar is probably not the source of all evil, which, it seems, is tantamount to boiling kittens alive.

It's interesting to see that this criticism comes almost exclusively from affluent, privileged commenters, who believe that they have discovered dietary secrets that can help free others from their shameful, fat bodies, offering a superior, skinny hand of friendship and guidance. They imagine that because they are thin, and they have managed to stay thin throughout their lives, their particular dietary beliefs must hold the key for all.

As they shop in Wholefoods after lunching at an exclusive organic café in Chelsea, they take the time to message me their wisdom, presumably before a leisurely afternoon at a posh spa. Clearly, the world is fat because people are not vegan. Or because 'the obese' all eat processed food. Or too much cake. Or sugar. Or carbs. Or maybe it's something to do with feminism. Or capitalism. Or agriculture. Or chemicals. It has never been possible to reply adequately to any of these criticisms in a tweet or even a blog post. This is not because I have no answers, but because the issues are complex and take some explaining. But with this book, I intend to answer them all.

Anyone who claims that they have a magic bullet* is either mistaken or lying to you. In researching for this book, I have spoken to some of the most brilliant and informed minds in the world, and none of them have The Answer. But by looking at everything we know and stepping back to understand how it fits together, I do think we can get closer to the truth.

A book that discusses problems without offering solutions would be of little use to the world. So as we progress, I shall start to shape a few ideas that might help us reach a better place. But be warned: *The Truth About Fat* is not a book that will attempt to sculpt your body. There is little chance of that because I have never met you, know nothing about your health, your mind or your relationship with food. All I can offer you is a greater understanding of why we get fat, a challenge to some popular myths about weight, and a little relief from the troubled relationship many of us have with our bodies. I will also shine a light upon some of the people exploiting our fears and misunderstandings, and try to make sure that you don't fall into their persuasive traps.

More than anything, I shall call for a world that tolerates and understands people, however they look, and attempt to strip away the last allowable prejudice of our age.

---

* That is, a solution, not a blender.

PART I

THE MODERN
EPIDEMIC

1

WHY DO WE GET FAT? PART I

I have had a half-decent career as a chef. As a slightly geeky science graduate with an interest in food, I entered the terrifying cauldron of a professional hotel kitchen in late 1994. Most of the other chefs thought that I would last a week or two, but through a combination of talent, bloody minded determination and the ability to out-drink most of my peers, I quickly rose through the ranks. Within twelve months, I was a sous chef in the patisserie section of the hotel. Within two years I was running the kitchen of a busy restaurant. Within five years I was in charge of twenty chefs. Within ten, I was the head development chef of one of the UK's largest food manufacturers, developing products eaten in millions of homes. My cooking career cost me injuries, stress, pain, and, at least twice, a genuine fear for my life.

In the past few years, I have somehow managed to develop a parallel career. I started a blog in 2016, and within a few months had an agent, a book deal, and was writing for a number of well-known publications in the UK and abroad. My first book sold well around the world, and was translated into fifteen languages. I now spend as much time writing as I do cooking, and it has transformed me in a wide range of positive ways. But writing my first book also meant that I did not have a day off for over six months, putting in sixteen-hour days in a cold, cramped, junk-filled spare room, distancing myself from my responsibilities as a husband and father.

Although many people have achieved far more significant things than I ever will, my career in the kitchen and my subsequent success as a writer are the most interesting things about me. I worked as hard as I could, and I am proud of what I have achieved. But apparently, there is something else about me far more worthy of note. I am frequently praised for it, especially by people I have just met. This week I received two random emails asking me about it, one from a follower in Thailand, and one from a new reader in the USA. It is literally being noticed around the world, and yet it is something that I have achieved with little to no conscious effort. I do not think it took a single moment of denial or sacrifice. No burns, no cuts, no stress or sleepless nights. Not even a cool scar to show for it.

What is this miraculous achievement, this secret that people are so desperate to know? 'How do you stay so thin?' People seem obsessed with the fact that even though I am a middle-aged chef who clearly loves food, vocally hates diets, embraces calorie laden 'junk foods' and occasionally eats breakfast at McDonald's, I somehow remain thin.

I say 'thin' because that is the language used by others when I am praised, but obviously this is fairly subjective. You are unlikely to find me stripped to the waist on the cover of *Men's Health* anytime soon, unless they have a dramatic change of editorial direction. If visible abs are the measure of worth that Instagram seems to think they are, I have a way to go. I own a pale, pallid torso, the muscle definition of a marshmallow, and a physique that the *Daily Mail* will never describe as 'enviable'. But I am not, technically speaking, overweight.

The last time I was measured, I was 75 kilograms (165 pounds). At a height of 1.82 metres, that sets my Body Mass Index (BMI) at $22.6kg/m^2$, which is considered to be nicely within the normal range. Although BMI is a crude and often misleading term, something that we shall look at in detail later on, it is widely accepted, and when it comes to analysing my body, reasonably accurate. I have never been called fat in any serious way, and do not consider myself to be so. Out of shape, perhaps. Fat, no.

THE MIRACLE OF CONSTANCY

I am forty-five years old and weigh roughly the same as I did when I was twenty-five. Although as a young man I was probably a little underweight, I never owned a set of scales, so cannot say for sure. I probably put on a few kilograms in my early twenties, but from twenty-five onwards my weight has been fairly constant. Despite this being beyond my conscious control, it is the thing for which I receive the most praise. The act of not getting fat. A great achievement and something of which I should be immensely proud.

And maybe I should be. After all, in the past twenty years I have probably burned around twenty million calories, and so, given that I have not gained any weight in that time, I must have absorbed twenty million calories from the food I have eaten. That sort of calorie intake required me to eat around 18 metric tonnes of food and drink, without any excess being laid down as bodily fat. This balancing act over such a sustained period is pretty remarkable. I am not sure whether or not I burned those calories according to how much I was eating, or if I managed to eat in relation to my energy needs, but either way it sounds like an impressive feat of calculation and control.

This is all the more remarkable because by any standards I have led a chaotic and poorly planned life when it comes to food. Some days, especially when working in kitchens, I skipped several meals in a row, staving off hunger by picking at various calorie-rich items throughout the day. I once went nearly a week consuming nothing but chicken skin, roast potatoes and Coca-Cola. Happy days.

There were many occasions when I had a huge takeaway, blowing out on pork ribs, pizza, curly fries or lamb jalfrezi, all washed down with family-sized bottles of cola. Often, I would eat until I felt I was going to burst, then somehow find room for ice cream or cake. But there were also times when I was ill, and had little to no appetite. I have had afternoons when I have run a half marathon. But I have also spent weekends on the couch watching the football, with potato waffles, cheese and cheap wine.

I have been drunk a number of times, consuming hundreds of calories in alcohol, before finishing the evening with a massive kebab. The next day I would nurse my hangover with a Big Mac, large fries and several energy drinks. Once or twice, lunch has been a twelve-course dinner at a fine dining restaurant. Other times, a family-sized pack of Maltesers eaten out of desperation while stuck in traffic.

And yet in all those years, I have somehow managed to maintain a constant weight, eating exactly the same number of calories as I have burned. Well done me. I'm a living, breathing miracle. My judgement of how much energy all those foodstuffs contain, all with different calorie densities, nutrient profiles and palatability, must be exquisite, especially given most calorie labels on foods are only around 90 per cent accurate. I must be capable of the most complex nutrient analysis imaginable, balancing this against subtle variations in my energy expenditure. And I have maintained this careful balancing act extraordinarily well over the long term.

THE IMAGINARY WORLD OF ANGRY CHEF

Imagine for a moment that my judgement was slightly out. Let us say that the day after my twenty-fifth birthday, I accidentally ate twenty calories more than I burned off. Just twenty calories, about the same amount that you might find in four olives. My body would not mind since it could easily store those twenty calories as fat. Now imagine that this slight miscalculation occurred on a regular basis every single day, for the entire twenty-year period up to now. Using the common estimate that 3500 calories is equivalent to around 500 grams of fat, over the first year, that excess energy would add up to around a single kilogram of weight gain. Not so bad. But if this continued, by the time I reached my forty-fifth birthday, instead of weighing in at 75 kilograms, I would be 115 kilograms (254 pounds) with a BMI of just under 35. I would be classified as obese, and it is highly likely that my life experiences would have been vastly different. My career prospects would have changed, the way people react to me would be completely altered, and the comfort with which I navigate the world would be greatly impaired. Katie Hopkins and Milo Yiannopoulos would think that I was disgusting, and never consider sleeping with me (every cloud). And all because of a single extra olive at each meal. And one before bedtime.

Now imagine that the difference was fifty calories a day, again a fairly insignificant amount of food. This is equivalent to less than half a tablespoon of olive oil, or a fifth of a single Dairy Milk chocolate bar. Over the three meals I eat most days, that amount of food would be unnoticeable for anyone living outside of a nutrition laboratory. Yet over twenty years, it would account for me gaining an additional 104 kilograms, meaning that I would be weighing in at 179 kilograms (395 pounds) with a BMI of 54. My life chances would be dramatically altered. Children would point and laugh at me in the street. The prime minister would make speeches about how people like me were ruining the country. Channel Five would offer me my own reality series. *Too Fat to Cook – The 30 Stone Chef on Benefits.*

One of the key things to understand about weight gain is that it rarely happens quickly. Most body fat is gained over years, not months, and for people to gain weight over that sort of timescale, it does not require a huge increase in calorie intake or drop in expenditure. The differences are likely to be unnoticeable to anyone not carefully weighing and measuring everything they consume. Yet our society and our media observe obese people and assume that their weight has a direct connection to food binges. And when people look at a middle-aged chef who is thin but does not diet, they assume he has some magical secret to share with the world.

Study after study has shown that, under normal circumstances, the majority of weight gain occurs slowly. The daily difference between calories consumed and expended, the famous 'energy in – energy out' equation, is generally very small. Even my four olives might be an overestimate of the average, with observed annual weight increases in populations likely to be accounted for by a difference of around nine calories per person per day. That's the difference between choosing to stand or walk on an escalator.[1]

I have controlled my weight over the years, but this has not been through any degree of *self*-control. I am often ill-disciplined, and frequently give in to desires and cravings. Although I enjoy exercise at times, I can be extremely lazy. The only reason I can offer for my thinness is pure luck. I have led a privileged life with a beneficial combination of good genes and a helpful environment, and this has resulted in me not getting fat. And in a world where thinness is seen as a proxy for moral superiority, it has handed me many life advantages.

The fact that my eating habits do not hide some hidden weight loss secret should not be too surprising. To understand the reasons why I am thin, it makes little sense to look in detail at my diet. And yet in order to understand why people are fat, the focus always falls upon what they eat. But what if the reasons had less to do with food intake than we think? What if large numbers of people get fat, even though they eat all the 'right' things? And what if others stay thin while eating all the wrong ones? Doesn't that make it illogical to blame one and praise the other?

In fact, the diets of most overweight and obese people, especially children, seem to differ little from those of supposedly normal weight.[2 3 4] Perhaps even more remarkably, despite the ubiquitous belief that people are fat because of what they eat, in 150 years of nutrition research no one has managed to establish a strong link between overeating, diet composition and obesity.[5] In the UK, as rates of obesity have increased, dietary surveys have actually shown decreases in consumption of sugar, fat, carbohydrates and total calories.[6] In fact, the only thing we are eating more of seems to be fruit and vegetables.

As we shall discover later on in the book, diet is an extremely weak predictor of weight gain, especially when compared against many other more powerful factors with an influence over bodily fat. Gareth Leng is a professor of endocrinology at the University of Edinburgh who has little time for anyone claiming that obesity is caused by a lack of will-power. On the link between diet and obesity, he told me:

> There is lots of attention on diet, I guess because it makes folk sense, but there is plenty of other stuff that has changed. It might be partly true that changes in diet have had an effect, but it's dangerous to mistake a plausible explanation for a valid one. There is very little evidence to show that the diets of obese people are different to people of normal weight, and diet generally is a very weak predictor of whether someone will become obese, far less than some other factors. Certainly, around the world there is a link between the level of food production and obesity rates, but if you compare developed countries, the correlation becomes very weak. Obesity is not a lifestyle choice. It is a multifactorial disease that is often a dysfunction in the hypothalamus.

This inconvenient fact is widely ignored not only within the media and the diet industry, but among many academics and public health professionals, all convinced that controlling and changing people's dietary behaviours is the key to helping them achieve sustained weight loss. Whether they claim that this should be achieved by shame, stigma, education, cookery lessons or a restructuring of the environment, the assumption is the same: poor food choices made you fat, and better ones will make you thin. Sorry, but it's not that simple.

WHY DON'T WE ALL GET FAT?

It is easy to postulate that evolutionary pressures and a competition for resources might keep wild creatures lean and hungry, and this could easily have kept a check on body weight throughout our evolutionary history. Nature is a cruel mistress, and likes to keep her charges in a state of near starvation, so the general leanness of wild populations is only natural. But when the majority of these pressures are removed, as they have been for those of us fortunate enough to live in developed economies, we are suddenly allowed free, almost unfettered access to food. And as we get richer, our agricultural systems and food supply chains become more efficient and robust, leading food to become a smaller and smaller proportion of our monetary expenditure. At this point, almost all of us have the freedom to eat well beyond our bodily requirements, with the tendency to lay down any excess calories as fat.

In this world of plenty, the fact that many of us manage to control our size with such extraordinary accuracy is unlikely to be accounted for by conscious effort alone. It seems incredibly unlikely that those who are not overweight are engaged in some superhuman feat of self-denial. It seems equally unlikely that our collective willpower has been gradually running out for the last forty years as obesity has grown into a so-called epidemic. Or do we think that people have stopped caring about their health over this time, which seems strange given how many have stopped smoking over the same period?

The most curious thing about body weight is not that some people have been getting fatter over the past few years. It is that, with the free access to food that our modern society enjoys, anybody manages to control their weight at all.

WHAT'S THE POINT OF FAT?

Our depictions of fat, and our general feeling about it as a substance, are in some ways indicative of the problems surrounding it. We see fat as a useless, inert jelly and create insults around it that imply sloth, inaction and ugliness. Yet fat is far from useless, and although lean creatures dominate the natural world that we regularly see, in many others, especially marine mammals, adipose tissue represents an essential requirement for life. Fat can insulate and protect us, and the development of subcutaneous* fat in humans is unique among large apes. Some think that it is what enabled early humans to lose their

* Situated under ('sub') the skin ('cutis').

body hair, and perhaps even facilitated them walking upright and developing many other uniquely human characteristics.[7] Fat, and the ability it gave us to store large amounts of energy, made us highly adaptable to variable conditions of food availability, temperature and seasonality, helping us colonise such a large proportion of the earth's surface in a relatively short time.

As we learn more about fat, and the cells in which it is stored, the idea of it as an inert, inactive substance is being revealed as far from the truth. Fat has a powerful, vital effect on the body. It is responsible for many signalling pathways that control important functions and even our behaviours. Although we live in a society at war with bodily fat, and it can be harmful to store too much of the stuff, it is important to remember that it is vital in ways that science is only just beginning to understand. One particularly relevant recent finding is that one of fat's main functions involves the complex signalling pathways that keep our body weight in balance. Although we spend a great deal of time, effort and money trying to control fat, it seems that it is very much in control of us.

WHO'S CONTROLLING WHO?

Right up to the early 1990s, when I was studying biochemistry at university, the prevailing thoughts about appetite control were that it just happened. People ate when they were hungry, in response to an empty stomach, and somehow the million plus calories they consumed annually were burned off through energy expenditure. Any weight gain over this time was a simple case of an imbalance in the 'energy in – energy out' equation, with excess calorie intake being converted into, and stored as, fat. Regulation of appetite was certainly not a significant area of study at the time, and the folk belief that self-control accounted for people staying roughly the same weight was taken for granted by most of the scientific community.

But not by everyone. As far back as 1969, the physiologist G.R. Hervey published a paper suggesting that our extraordinary long-term regulation of body weight was more than just a coincidence.[8] The laboratory rats he studied regulated their body weight extremely well, even with free access to as much food as they desired. When their food was diluted with an inert, calorie-free substance, they simply ate more of it, adjusting their portions incredibly accurately to prevent any weight loss. Sceptical that rats knew how to count calories, Hervey suggested that their fat stores were somehow signalling to their brains. He particularly focused on the hypothalamus, as rats in which this area of the brain had been altered sometimes became obese through over-eating. In subsequent years, experiments demonstrated the same effects in humans, with subjects simply consuming more of the diluted food over time to avoid a calorie deficit.[9] As with rats, people show remarkable accuracy in doing so, with little evidence of any conscious control. But Hervey's research was largely ignored, and the prevailing view of dietary self-control continued to dominate for some time, showing an impressive confidence in the desire of rodents to keep themselves trim despite not being able to turn the pages of *Vogue*.

A few years after Hervey's experiments, some interest did start to develop in the heritability of obesity, with studies of twins indicating there might be a significant genetic component (we shall look at this later). For this reason, a number of researchers spent time attempting to breed overweight rats and mice for study, to see if any insights could be garnered into the genetic factors at play. The two most interesting types produced were the 'ob/ob mouse' and the 'db/db mouse', both of which became extremely fat through overeating.

Both these mice had a single defective gene, and it was only homozygous* versions of them that gained large amounts of weight. As well as getting extremely fat, the db/db mouse often developed type 2 diabetes, something that interested the researcher Douglas Coleman, and prompted some new experiments. Coleman linked the blood supply of a db/db mouse to that of a normal mouse, and, to his surprise, found that the normal mouse stopped eating almost completely. He concluded that some sort of substance in the blood of the db/db mouse was causing the normal mouse to believe that it had enough fat in stor-age and did not need to eat, but for some reason that substance was not having any effect on the appetite of the db/db mouse.

Coleman then paired db/db mice with ob/ob mice and found that the ob/ob mice also stopped eating and lost their excess weight. Perhaps, Coleman thought, the ob/ob mice couldn't produce the chemical that stopped them from eating, and the db/db mice couldn't

---

* We have two copies of each of our genes, split across paired chromosomes. When both copies of the gene in a pair are the same, the gene is 'homozygous'.

respond to it. Whatever it was, it clearly showed that there were aspects of appetite that were beyond conscious control, at least in mice. Given the blood of a fatter mouse, a normal mouse would happily starve itself to death, despite free access to food. It was fooled by the new blood into thinking that it had no need to eat.

Sadly, like Hervey's before him, Coleman's work was largely ignored by the scientific community at the time. He was later quoted as saying that 'despite these clear results, many in the obesity field maintained the dogma that obesity is entirely behavioural, not physiological'.

It seemed that even though there was new and compelling evidence, the view of fatness as a product of moral degeneracy was just too hard to break down, even for the supposedly objective scientific community.

But not everyone ignored this work. In 1994, having been inspired by Coleman's experiments and using a new technique called positional cloning, a team at Rockefeller University, under the direction of Professor Jeffrey Friedman, identified the gene that caused the ob/ob mutation in mice, something famously depicted in a picture on the front cover of the prestigious science periodical *Nature*. On one side of a balance sat two normal sized mice, on the other an ob/ob mouse weighing them down, over twice their body weight. The paper described the discovery of the very factor that Coleman had imagined, a hormone that the researchers named leptin. It was the first new hormone to be discovered in fifty years and, at last, the evidence that appetite was regulated by factors beyond sheer willpower was too great to ignore.

THE OBESITY HORMONE?

Leptin is produced by fat cells ('adipocytes') and acts along the lines that Hervey had guessed at in the late 1960s. He was also correct about the role of the hypothalamus. Leptin signals to that part of the brain, which then activates complex regulatory mechanisms to control the appetite. When there is plenty of fat in storage, adipocytes produce lots of leptin, and so decrease appetite. As fat stores start to be used up, leptin production drops, and the hypothalamus sends out signals to start eating more.

The poor ob/ob mouse cannot produce leptin and so spends its life always hungry. Despite the mouse growing rapidly to a considerable size, its hypothalamus is constantly telling its body that it has no fat in storage. It unconsciously thinks that it is on the brink of starvation, driving it to eat.

As for the db/db mouse, it produces plenty of leptin, yet its particular mutation means that it has no receptors for the hormone, so cannot detect its presence. While its fat cells are screaming out that they have more than enough energy in storage, its brain cannot pick up the message, and so it too eats ravenously, as if it was starving. When Coleman connected the circulation of a db/db mouse to that of a normal mouse, high levels of leptin flowing through the blood signalled that it was overweight and it stopped eating. It seems that in leptin, science had discovered a way to shut off the appetite. At least in mice.

Not surprisingly, attention quickly turned to whether or not a similar system might exist in humans, and if this might be used to regulate people's eating. Friedman's team identified the same gene in humans, and, fairly soon, a small number of individuals had been discovered who carried a similar mutation to the one found in the ob/ob mice, meaning that they too were unable to produce leptin.

This rare mutation, usually only found in the children of first cousins, had a profound effect on the lives of those unfortunate enough to have it. It results in severe obesity from early childhood, caused by excessive appetite and non-discriminatory eating habits, a condition known as hyperphagia. As children, sufferers will fight others for food, eat in secret, hoard and binge, and readily eat items that most of us would find unpalatable, such as uncooked fish fingers straight from the freezer. Their lives are consumed by a constant and extraordinary hunger, with parents required to keep fridges and food cupboards locked. They gain weight exceptionally quickly, and even under strict dietary controls will struggle to lose any bodily fat. They also have a low functioning immune system and rarely enter puberty, usually remaining infertile into adulthood. Leptin is a powerful hormone indeed, and a lack of it makes for a thoroughly miserable life.

But remarkably, once a leptin-based hormone treatment had been developed, these individuals returned to a normal weight incredibly quickly. They would not just stop gaining pounds – they would lose enormous amounts of fat, just as the ob/ob mouse had done when leptin was flowing in its blood. Their appetite came under control, their immune system started to function properly, and they would go through puberty in the normal way. In the last few years, the first sufferer in history gave birth to a child, something that would never have been possible without leptin.

HOPES FOR A WONDER DRUG

For the researchers involved, it must have seemed that not only had they managed to alleviate the symptoms of a rare genetic illness, but also that they might just have cracked the problem of obesity. Here was a safe and effective hormone that could be given to obese people, after which they would return to a normal weight without being plagued by gnawing and miserable hunger. If a lack of leptin was causing severe obesity in this small group, perhaps less serious cases were caused by slight decreases in the production of the same hormone. Unfortunately, things were not quite that simple, and leptin never proved to be the obesity curing wonder drug it once promised to be.

It is important to remember the leptin pathway exists to guard against starvation, not to prevent us from becoming fat. With no leptin present in their blood, the brains of those with hyperphagia unconsciously feel that they are approaching death, and so their bodies enact emergency procedures. A lack of the hormone creates the strongest possible drivers for food, and also acts to shut down any unnecessary energy expenditure. The immune system is largely discounted because it is just a theoretical insurance policy against future infection, something of little use if you are starving to death. Puberty and fertility are also unnecessary concerns when food is scarce, and becoming pregnant potentially catastrophic at such a time. There is good evidence that a lack of leptin also causes certain aspects of cognitive function to be altered in order to preserve calories, as well as generally decreasing all energy expenditure. This can lead to lethargy, a lack of normal growth, loss of hair and nails, and reductions in body temperature. Everything possible is done to decrease the burning of unnecessary calories, with all energy channelled into a compulsive drive for sustenance, creating an insatiable and near primal appetite. A lack of leptin reveals one of our most important instincts, the desire to stay alive at all costs. It is telling that a lack of the hormone, rather than its presence, produces these powerful effects, so conserving even the energy required to make it.

If you do not have this mutation, you should only ever experience these effects during starvation. Once we are fed and manage to store a little bit of energy as fat, we will have some leptin flowing through our veins once again, and our body will know that we are safe from immediate danger. We then become free to do all the things that people like to do beyond seeking food, using our energy to find a mate, to think, to plan, to create, and to live our lives. In short, leptin, and the ability to respond to it, allows us to be human. It also allows mice to be mice, and rats to be rats. Without it, we are all just empty vessels, searching desperately for food in a bid to survive. For any species to survive, these emergency signals have to be uniquely powerful, and incredibly hard to override.

When it came to using leptin as a drug treatment, the problem was that, in general, our billions of fat cells seem to have little problem producing it. In fact, the majority of obese people have extremely high levels of leptin in their bloodstream already. For that reason, administering more leptin has little effect on body weight or appetite. Sadly for the researchers, most obese people have little in common with ob/ob mice.

But there was another type of mouse that got fat. In recent years it has emerged that human obesity might have more connection to db/db mice and their inability to respond to the hormone. In many obese people, although they do not lack receptors for leptin in the way that db/db mice do, it seems that they might have differing levels of sensitivity to the hormone. Although obese people usually have high levels of leptin in their bloodstream, many seem far less responsive to it than people of normal weight, something that is highly likely to be a driver of appetite and weight gain. It has been estimated that between 10 and 15 per cent of severely obese people have gene defects involved in the circuitry that responds to leptin.[10] There are other reasons why someone's ability to respond to leptin might change, some of which we shall explore later on, but for now you just need to remember that much of our appetite is guided by the same unconscious, ancient instincts that can cause such extreme hyperphagia in leptin-deficient children. An insensitivity to the hormone has the potential to cause a drop in metabolism, a decline in sexual desire, a lowered immune system and severe alterations of mood. It can also awaken a primal desire for food, something that has developed over millions of years to prevent us from dying, and perhaps the most powerful instinct that we possess. Seen in this context, losing weight is not always as simple as laying off the cheese. When obesity is framed as a genetic blueprint to be sensitive to certain hormones, how much blame can we attach to those it affects? And what do we really think our constant shame and stigmatisation are likely to achieve?

We are consistently fed the story that thinness can only be achieved through discipline and self-control, and fatness only occurs when you cave in, yet this is far from the whole truth. Although we might like to believe that everything is a matter of free will, our hormones are hugely powerful and staggeringly complex, producing integrated systems that control every aspect of our lives.

Although leptin may be the most significant appetite hormone, it is certainly not the only factor that affects our weight. As you'll see in the next two chapters, the real reasons why people gain weight, and the reasons that many of us can't shift it, are not what you might expect.

2

WHY DO WE GET FAT? PART II

The discovery of leptin in 1994 led to a massive increase in research to find out what else our fat cells do and what is going on in the gut. We all sometimes 'feel' hungry, an unpleasant physical sensation of emptiness, and it is no surprise to find that our guts also exert a powerful hormonal influence on the body.

In 1999, the discovery of a hormone named ghrelin was reported by a group of researchers in Japan, led by Masayasu Kojima.[1] Ghrelin is an extremely powerful signaller, and increases the drive for food intake rather than suppressing it. When our stomach is empty it produces ghrelin, when it fills up it stops producing as much. Ghrelin also signals to the hypothalamus, activating hunger pathways that drive us towards food intake and decrease unnecessary energy expenditure. Ghrelin turns out to be responsible for much of our short-term drive for food, and can be an extremely powerful motivator. Clearly, how much ghrelin we produce, how well we respond to it, and how quickly we start producing it again after we have eaten, will have a huge impact on the amount that we eat.

Both ghrelin and leptin influence the same systems in different ways, carefully balancing short-term hunger with long-term fat reserves in order to keep us satiated and in balance. We need a wide range of micronutrients in order to stay healthy, so we still might need food even when our fat reserves are plentiful. There is now good evidence that these two hormones act in synergy in order to keep us the same weight.[2]

Many other hormones produced by the gut have since been discovered, and there are powerful neural signals running straight from the gut to the brain, through something known as the vagus nerve. Amazingly, our gut can detect the nutritional composition of the foods we eat, activating specific metabolic pathways to aid digestion. Different foods will have varying impacts on the length of time it takes before we are hungry again, thus controlling our calorie intake. A variety of hormones and other signals will activate desire for particular food types.

We have all craved carbohydrates, especially after deciding to stop eating chips last week. Yet like many things related to appetite, this too was considered to be purely psychological until relatively recently. In rats, ghrelin has now been shown to increase the enjoyment of certain foods, particularly energy dense foods high in fat.[3] You know how the first bite of a chocolate cake is so much more enjoyable than the last? We used to think that this was purely driven by our psychology and associations, but it seems that some of this feeling is actually coming from our gut hormones.

There are now many other hormones known to act upon our appetite, signalling to the hypothalamus in order to control our behaviours and metabolism. In addition to ghrelin and leptin, there are roles for insulin, glucagon, CCK, PYY, GLP-1, GIP, OXM, peptide YY and PAMP, which, while they might sound like a particularly rubbish superhero franchise, all affect how we eat and digest. Adipose cells, once thought to be little more than inert fat stores, are now known to produce many different active substances, with impacts all around the body that we are only just beginning to understand.[4]

As we shall see later on in Chapter 11, exercise produces a number of hormones that stimulate future appetite, powerfully driving calorie compensation in order to prevent weight loss (our bodies really don't like losing weight, as we shall discover in the coming chapters). Perhaps even more incredibly, recent experiments on animals have indicated that there are mechanisms that monitor the physical load on our joints and muscles, releasing factors that regulate appetite in response to changing body weight.[5] This raises the possibility that we have inbuilt weighing scales, and a system that stops us from getting too heavy, regardless of how fat we are. Perhaps the reason I never needed a set of scales as a young man was because I already had one built into my body, signalling to my brain every time I gained or lost a pound. The researchers even suggest that these mechanisms might explain why people who spend a lot of time sitting are more likely to become obese. Perhaps remaining in a sedentary position, without load on our joints and bones, leads our body to think it is lighter than it really is. And when our body thinks it needs to gain weight, there is little our conscious selves can do to stop it.

Coming from a starting point of general scientific disinterest, the discovery of a single new hormone in 1994 has catapulted the field of appetite regulation into an area of staggering complexity. There are a great many important behavioural cues and associations in play, including some that might encourage us to consume highly rewarding and calorie dense food when it becomes available, even when we are not hungry or starving. Eating egg yolk or animal fat might have provided an evolutionary advantage at a time when these types of food were scarce, precious and in danger of rapidly spoiling. But you can see how it might present a problem today.

When we eat, a huge number of pleasure and reward systems are activated, and powerful psychological associations can develop with particular items. Our olfactory senses give us a crude indication of nutritional content and safety, often offering up the greatest sensory rewards for the most calorie dense items, most likely to be concentrated sources of fat, starch, sugar or protein. Certain combinations are particularly rewarding and pleasurable, perhaps even addictive (although we should be very careful not to associate pleasure with sin).

After swallowing, perhaps remarkably, we are not done with tasting. Our guts contain taste receptors, as well as sensors to analyse the amount of food eaten, and even the chemicals it contains.[6] The gut sends huge numbers of hormonal signals, as well as massive amounts of information through the vagus nerve. The importance of these signals in regulating our appetite has been shown in experiments where the vagus nerve signals are disrupted, something that can lead to larger meal sizes and overconsumption.[7] In fact, this 'enteric nervous system' is so sophisticated and complex that it is often referred to as our 'second brain'.

As well as reading signals from our gut and our fat cells, the brain also receives signals from the liver and the pancreas, which respond to the availability of different nutrients. All these signals affect how we eat, and serve to create positive and negative associations with different foods. At an unconscious level, masses of integrated systems combine, favouring the most rewarding foods. Consciously, this results in us enjoying cake, biscuits and any other foods that have given us pleasure and nutritional rewards in the past. These appetite and reward pathways will motivate us to eat bacon, bread, eggs, cheese, or whatever our body feels will most closely match its requirements, and will also meet our important emotional and psychological needs. Although we make conscious food choices, many of our desires and cravings are strongly linked to many other unconscious reward and motivation pathways. For this reason, pleasure, stress, emotion, tiredness, relationships, sleep patterns, physical exertion and numerous other factors are involved in the control of our food choices, and can also have a significant effect on how the food we eat is processed and stored by our body.

It is hardly surprising that this system is so complex. After all, it has developed to guide and control food intake for many millions of years of evolution. The systems that regulate food intake are similar among many different species of mammals and birds. Such staggering complexity is hardly surprising when you consider exactly what it is these processes are trying to achieve. Animals can migrate vast distances, often expending extraordinary amounts of energy without taking on sustenance. Some hibernate, and most will experience times of plenty as well as periods when little or no food is available. Throughout their evolutionary history, humans especially have had to adapt to widely varying conditions and food availability around the world, colonising environments of colossal diversity, from the plant-free, meat-rich regions of the high Arctic, to the lush green rainforests of South America. We have evolved a massive brain with a huge energy requirement, yet with a system so flexible it can run on completely different fuels for long periods with little impact on performance. Our appetite regulation needs to ensure that we get enough calories, but also the correct nutrients, in order to stay healthy. Some vitamins cannot be stored by our body, so we need to be strongly motivated to seek them out. We have to lay down fat as insulation and energy store, but also stay lean enough to hunt.

These processes have ensured that humans meet their nutritional requirements. Yet until the last 100 years, no one had even the slightest idea what those requirements were. For millions of years, across thousands of different species, the complex hormonal control of appetite has managed to keep creatures consuming the correct quantities of micro- and macronutrients required for life. It has done a remarkable job, and any idea that it might be simple is clearly absurd.

CALORIES OUT

Curiously, there is good evidence that for many people, an increase in body weight might not require an increase in calorie consumption at all. In most scientific studies it is difficult to determine how much people are eating, because people generally submit their own data and this isn't seen as reliable. (Have you ever had to tell your doctor how much you drink or smoke?) But in certain controlled environments, where sneaking in an unreported chocolate bar is unlikely, interesting insights can be gained. US prisons have long been fertile ground for nutrition research, with a number of prisoner-feeding experiments being conducted over the years. Overfeeding studies in the 1960s by the researcher Ethan Sims showed that subjects gain different amounts of weight despite eating identical amounts of food,[8] indicating that the 'calories out' side of the equation varies considerably between individuals, independent of exercise or other physical activity. It also showed different amounts of resistance to weight gain, as subjects' metabolic rates seemed to increase on over-eating, their bodies trying hard not to lay down too much of their excessive food intake as fat. The extent to which this happened varied significantly among individuals, despite the similarity of their diet and environment.

Obviously, these results are only curious if we assume that every-one reacts to food in exactly the same way, and although this principle underlies the majority of dietary guidelines, diet books, weight loss protocols and nutrition advice, experience tells us that it is a bit more complicated than that. The way we process food, burn or store calories and react to different nutritional intakes is likely to vary greatly, even between outwardly similar people. A diverse population would have a number of evolutionary advantages, with some individuals prioritising the effective storage of fat in order to survive and others burning most of their energy exploring. It would also make sense for us to be able to switch between those two states when we needed to.

But how is it that the amount of weight gained from identical food intake can vary among individuals? Even though hormones regulate our feeding, we only store calories when we eat more than we burn, right?

Our expenditure of calories can be broken down into four parts:

1. *Basal metabolism*, the amount of energy we need to stay alive.
2. *Thermogenesis*, the energy required to keep our body weight constant.
3. *The thermic effect of food*, the calories expended in processing and digesting food after eating.
4. *Physical activity*, the amount that we expend through exercise.

All these have their own pathways of regulation and control, and, as the prisoner studies show, can vary considerably among individuals.

STAYING ALIVE

When we consider the 'calories out' side of the equation, we almost always look at exercise as the most important factor, largely because it is the one area that we can easily change. But basal metabolism is by far the largest chunk of the energy we burn. We generally expend around half of our calories staying alive, and another 10 per cent staying the right temperature. Around 30 per cent goes towards physical activity, and the remaining 10 per cent on digesting food.[9] [10]

Looking back again to the twenty-five-year-old version of me, it is perfectly possible that a slightly altered version of that mixed up, tormented chef might have gained considerable weight while consuming exactly the same quantity of food and drink. As the majority of our calories are taken up by the effort required just to stay alive, an expenditure entirely beyond conscious control, it would not take a great deal of movement in basal metabolism to vastly change the amount of energy stored as fat. A hardly perceptible 5 per cent shift, easily possible given known genetic variations among people and well within the differences observed in the prisoner studies, would have led me to gain several kilograms of excess body fat every year. The alternative version of me would be left weighing around 225 kilograms (495 pounds), with a BMI of 68 on his forty-fifth birthday, if he lived that long.

If he did manage to stay alive, this particular Angry Chef would be classed as super-morbidly obese, and considered a disgusting moral failure, despite there being no difference between his eating habits and mine. Social media messages from Thailand and the USA would likely be abusive insults about his disgraceful, embarrassing fatness, rather than congratulations about his remarkable self-control. He would have lived a very different life, all without the slightest change in the food he had consumed.

Perhaps even more significantly, variations in basal metabolism do not just exist among individuals: our environment can greatly alter the rate at which we burn calories at rest, with the potential to hugely affect our weight. Often the reasons for these changes are surprising and counterintuitive.

THE BIGGEST LOSER

Kevin Hall is a US-based researcher at the prestigious National Institute of Diabetes and Digestive and Kidney Diseases. However, in the world of nutrition, he considers himself to be something of an outsider, largely because his PhD training was in physics rather than the biosciences. Despite this, in recent years, his work has garnered particular attention, perhaps because as an outsider he is less constrained by some of the orthodoxy that hampers so many academic fields.

His team is attempting to create mathematical models of human responses to nutritional intake, all based on existing scientific knowledge. They then attempt to test and validate the predictions made by these models by conducting human experiments to show just how complete our existing understanding is.

Perhaps revealing some of his physicist training, Kevin claims that 'I don't really understand a system unless I can build it for myself' and this approach has been very successful in pushing the boundaries of knowledge when it comes to nutrition science. The models he and his team create take into account all the known hormonal and neural responses to food that we have discussed and, despite the great complexity involved, they have produced some marked successes. Kevin told me:

> The purpose of creating a model is to see if it can make good predictions based on our existing knowledge, and I think our models have shown that we have a pretty good understanding of

how the body reacts to changing macronutrient* intakes, and the endocrine processes of shifting fuels, say from carbs to fat. The models have made good predictions about the certain outcomes, but it is far more interesting when the model predictions are wrong. Then we discover new things. You never really know when you are right, but you always know when you are wrong. And early in 2016, our model went spectacularly wrong.

The work that he is referring to was the end of a six-year study into contestants who had appeared on Series 8 of the popular US television show *The Biggest Loser*. In the show, obese and overweight contestants compete to lose the largest percentage of their body weight over a defined period. Contestants are given access to trainers, nutritionists and medical professionals, who devise comprehensive weight reduction programmes for them. Gradually contestants are whittled down based on the amount of weight they have lost, eventually competing for the prize of the 'Biggest Loser' in a grand finale.

With competition for huge prizes and the status of winning such a high-profile show, the weight loss that some contestants manage to sustain is quite extraordinary. It is not unknown for some to achieve a 10-kilogram loss in the first week, and continue dropping 4–5 kilograms every week. This is vastly more than health professionals recommend, requiring extraordinary willpower and potentially dangerous, starvation-level calorie restriction, combined with six hours a day of strenuous exercise. There have been a number of hospitalisations, and Ryan Benson, the winner of the first series, admitted to 'fasting and dehydrating myself to the point I was urinating blood'.

The Biggest Loser has a despicable fat-phobic premise, apparently designed to create disordered behaviours in its contestants. Still, if you wanted to perform a case study into the effects of rapid weight loss on people's metabolism, it is a great place to look. Kevin Hall's team studied the contestants when they were chosen, then at six weeks, thirty weeks and six years after the show began. After six years, most of the contestants had regained the weight they had lost, with four of them being heavier than when they started. The most

---

* These are the common substances from which we get the majority of our energy. The main ones are carbohydrates, fats and proteins. Water is not considered because it has no energy value. A case is sometimes made for alcohol, but it is hardly essential (at least not on weekdays).

surprising finding was the enormous change in each contestant's basal metabolic rate, something that hugely affects the number of calories needed daily in order to maintain a stable weight. On average, contestant's metabolic rates were 500 calories lower than before the show had started, even after six years. The winner, Danny Cahill, had lost an extraordinary 108 kilograms (239 pounds) during the show, and had regained around 45 kilograms (100 pounds) six years later. His metabolism had dropped by an extraordinary 800 calories per day. Leptin levels were also shown to be significantly reduced, perhaps giving a clue as to some of the mechanisms at play. As Kevin says:

> Here the predictions we made were very wrong. We understood and expected the drop in metabolism during weight loss. This is a normal metabolic adaptation that has been repeatedly observed, but we only predicted a short-term drop while the participants were actively losing weight. The metabolic rate was expected to recover when weight stabilised or partly regained, but it did not. The question is, is this a translatable phenomenon, does it occur in less extreme circumstances when people engage in less exercise and eat more calories?

Although the study does not provide a complete picture, it is fascinating because it provides a window into how powerfully our bodies can protect a particular weight, even when that weight is not thought optimal to our health. It also shows that there is a great deal that we do not yet know.

Perhaps because of its shocking, headline-grabbing findings, the study has been much reported in the media, especially in the US, where for some reason the show remains extremely popular. However, one point that was rarely picked up in the coverage gives another tantalising insight, suggesting that something even more mysterious might be going on. Although it might have been expected that when the metabolism slowed down (i.e. burned off less energy), it would be directly linked to people regaining weight, there was evidence that this was not always the case. In fact, Kevin says: 'There was no correlation between metabolic slowing at the end of the competition and subsequent weight regain. The people who had experienced the least weight

gain since the show actually continued to experience the greatest slowing of metabolism.'

Despite this work, the potentially huge factor of our variable basal
metabolism is rarely considered. For a start, it has been difficult to get
good quality data. Information on how many calories someone is burning is hard to obtain, requiring either careful measurements using
radioactive isotopes or shutting volunteers in a sealed chamber that
carefully monitors their intake of oxygen and exhalation of carbon diox-
ide. These techniques allow total energy expenditure to be calculated,
but they are complex, difficult and expensive, especially for any large-
scale studies. So even though metabolic rates are an important part of
the picture, academic studies inevitably focus on things that are easier
to measure, such as exercise calories, food intake, weight, height and
BMI. This skews what we know about obesity, which leads us to focus
on measures such as calorie reduction and increased activity.

The *Biggest Loser* study probably poses more questions than it
answers, but it undermines the idea that food consumption and exercise are the only factors that influence weight gain, hinting at a far
more complex system of control. An 800-calorie metabolic difference
is equivalent to running six miles every single day. If you weren't exercising and wanted to lose weight, you would have to push your calorie
intake to near starvation levels, no matter how heavy you were to start
with. In such circumstances, it would be a challenge to get adequate
nutrition into your body, probably requiring extensive supplementation
and professional dietary support in order to avoid becoming ill from
malnutrition.

And if that is not bad enough, just think exactly how that 800-calo-
rie reduction in basal metabolism might be achieved by the body. For
most people, 800 calories represents over half their basal energy
expenditure, which is set at the level it is for good reason. The body
does not burn fuel for fun, and the savings have to come from somewhere. Remember how leptin-deficient children reduce unnecessary
expenditure of calories? It happens by shutting down the immune
system and menstrual cycles, dropping body temperature and dimming
non-essential cognitive functions. That is what your body would do to
you, using exactly the same pathways downstream of the hypothalamus, enacting profound changes in a perceived response to life threatening starvation. You would become sick, tired and cold, but most of all
you would be hungry. A constant, gnawing hunger, a primal instinct

designed to protect you. In order to stay the same weight, you would have to put up with this for years, constantly starving your body of what it desired. If you returned to eating normally for just a short time, you would very rapidly become obese. The willpower of anyone who does not gain weight in these circumstances must be extraordinary. I am certain that I would never manage it myself.

All this is worth remembering next time you hear someone say, 'It's easy, just eat less and move more'. If you are still holding onto the idea that obesity is a matter of either willpower or gluttony, then be warned. In the next chapter, things are going to get even worse.

3

WHY DON'T WE
JUST EAT LESS?

Britain is now the fattest country in Europe, and gaining fast on the US. It's not genetic, it's gluttony. The solution is to exercise more and eat less. All it takes is a little will power.

Richard Littlejohn, Daily Mail

To say that obesity is caused by merely consuming too many calories is like saying that the only cause of the American Revolution was the Boston Tea Party.

Adelle Davis

Diet fads are for the birds, if you don't like birds.

Ancel Keys

IT'S EASY. ISN'T IT?

Despite all these effects, weight loss is easy, right? Well, perhaps not easy, but certainly simple. You still need to eat a bit less and move a bit more in order to get thin, and the likelihood is that you are going to have to consume fewer calories. Shovel a bit less food into your pie hole. Maybe get some exercise now and again. But it takes a lot of exercise, so mostly just eat less.

If the energy-balance model is true, and let's face it, it is based on some fairly irrefutable laws of thermodynamics, then so is the simple adage: to lose weight, you need to eat less and move more. It is so simple that it dates back to the fourth century BCE, indeed to Hippocrates, the father of modern medicine. Quite remarkably, this has remained the cornerstone of weight-loss advice ever since. Although we have learned that our hormonal responses make this advice difficult to follow at times, it is surely still true. Certainly, if you read the comments on every single obesity related story in the media, there will be a thousand people telling you just how simple and irrefutable this ancient piece of medical advice is.

Maybe this common-sense reasoning is correct (although to be honest, if it is, I'll probably have to end the book right here). After all, despite the known complexities of hormonal weight regulation, calorie-reduced diets are still the primary treatments prescribed for the majority of overweight and obese people. Visit your doctor tomorrow, and that is what you will be told, and you will likely be given various perfectly sensible sounding strategies to achieve it. In fact, if you are fat, whenever you visit a doctor, whatever that might be for, it is quite likely that you will be told to eat less and move more.

This anecdotally results in fat people being told to lose weight whenever they drop in to the doctor with an ingrowing toenail, a sore hand or a severed foot. The management of people's weight is considered such an important health priority that dietary interventions are often suggested to fat people, even when they haven't been asked for and aren't relevant to the condition being treated. Despite the likelihood that this sort of unwanted judgement could have a stigmatising effect, perhaps even preventing overweight and obese people engaging with healthcare at all, it is still part of the guidelines for primary healthcare professionals in the UK. And the suggested intervention is usually the implementation of a calorie-restricted diet. The UK's National Institute for Health and Care Excellence (NICE) guidelines for weight management recommend referring obese patients to diet programmes such as Weight Watchers, Slimming World or Rosemary Conley, based on these being proven to be effective over a twelve-month period.

Obviously it is not just at the doctor's surgery and in hospitals where this advice dominates the weight-loss agenda. The media is awash with diets and various fat-busting strategies, all based on the premise that consuming fewer calories is the best way to shift excess pounds. I have spent a long time looking at diets, and, despite outward differences, they are all essentially the same: restrict a food or foods in order to take on less energy than is being burned. Most create a complex and fairly arbitrary series of rules, designed to make the follower eat less food. Some hope to achieve compliance by the creation of fear, deeming certain foods as acidic, unclean or toxic. Most of the rules created are based on pseudoscience, often using the quite absurd narrative that we are not 'designed' to consume commonplace, calorific nutrients. Many advocate the cutting out of carbohydrates, claiming sugar, starch, grains or gluten possess a mysterious toxicity that makes us fat. But they all maintain that weight loss is simple, and that all you need to do is restrict the calories you are eating. The most honest will admit that this is likely to be very hard work, requiring a great deal of willpower and self-control. But all profess to be straightforward. Just eat less, move more.

This simple equation implies personal responsibility: if fat people just ate less, they'd lose weight and we wouldn't be burdened by the ugly sight of their fat faces, their bellies wobbling around in front of us, their sickening folds of skin, an unsightly, smelly stain on our everyday lives. I mean, obviously, most people don't say that out loud, but it is pretty clear that it is what they're thinking. Because diet culture depends on us being disgusted and repelled by fat people. Without that, we'd all just tell people to fuck off and mind their own business.

IS IT REALLY THAT SIMPLE?

In the last chapter we learned that hormonal regulation can be very powerful, dramatically changing our metabolism if it needs to, albeit in an extreme situation. But how does that affect people on normal weight loss diets? Well, as it turns out, quite a lot.

Regulation of our body weight has been vital for our survival as a species, and evolved to be a carefully controlled process, unlikely to respond very much to small day-to-day changes in food intake. Although this might seem to be at odds with the 'calorie in – calorie out' idea, it really isn't. But it does require us to think a little deeper about how the 'calorie out' part might change over time, and in response to different stimuli. Whenever people restrict their calorie intake, powerful forces come into play, because the one thing the human body does not like is being deprived of energy. To assume that losing weight is just a matter of cutting back for a while is to assume that there are no processes in place that will try to stop weight loss from occurring.

As soon as calorie intake is restricted to levels below the daily requirement, appetite- and metabolism-regulating hormones come into play. Calorie-restricted weight loss has been shown to alter the levels of many of these hormones for up to a year after dieting,[1] and not just under the sort of extreme conditions studied in *The Biggest Loser*. These changes have been shown to stimulate appetite, increase the desire and perceived rewards for high-calorie foods, reduce the metabolic rate and decrease bodily temperature, so making it far harder for dieters to lose weight, and likely to make them feel pretty unpleasant while doing so. If you cut a few hundred calories from your daily intake as most diets recommend, the likelihood is that after initial weight loss, you will pretty soon be losing nothing at all. A study at Columbia University showed that subjects who lost 10 per cent of their body weight had basal metabolic rates between 250 and 400 calories lower than they should have had, meaning that they would have to eat considerably restricted diets just to maintain a constant weight.[2]

A study of young women who lost weight after being placed into short-term calorie restriction showed that they regained that weight rapidly after being allowed to eat unrestricted amounts of food.[3] Although this was to be expected, importantly the weight gain occurred without the women eating any more than they had been previously, indicating that it was largely caused by changes in their metabolic rate, rather than increased appetite.

Numerous studies have shown that altering the energy content of meals without informing subjects results in them unconsciously compensating for those reduced calories elsewhere. This can occur either through an increased intake of other foods, a drop in basal metabolic rate, or other reductions in calorie expenditure. Other studies have shown that even though a restriction of calorie intake might create a deficit and weight loss in the very short term, the body tends to even out these changes over a four- to five-day period, altering energy expenditure or changing appetite accordingly. So even if you eat less today, your body will make up for it later in the week. And when you exercise, your body will compensate for several days until it has got back the calories burned.[4]

BUT DIETS WORK, RIGHT? I MEAN, DORIS AT WORK LOST WEIGHT LAST YEAR.

Even if these difficulties can be overcome, and clearly it is possible for this to happen, anyone achieving short-term weight loss by calorie restriction is likely to encounter problems. As we have seen, hormonal changes can last for several years after dieting, and trying to fight them for that long is overwhelming for most people.

Although studies into the long-term success of diets are hard to conduct, the picture is really not great for this regularly prescribed medical intervention. It seems that each body has a 'set point weight' that it likes to defend, and whatever happens, it works hard to get itself back to that particular weight, whatever it might be. Sadly, it is the body that defines this point, not the dieter, and if there is a difference in opinion as to where it should be, a battle will ensue. And the majority of the time, it is the body that wins.

In reality, this point has the capacity to change over time, sometimes dramatically. This should not be too surprising, as many animals also have a variable set point weight. As Gareth Leng, the endocrinology professor, explained to me, there is plenty of evidence that set points can alter over time in response to environmental stimulus:

> Hamsters have a variable set point in response to photoperiod, so making their weight seasonably variable. If you artificially alter their weight away from where it should be during the year, it will return back to where it should be when you return them to a normal diet. Similarly, sheep have seasonal differences in body weight, and as they are pretty much always eating as much as they can, those changes are driven by altering their metabolic rate, not their appetite. All mammals, including humans, will alter their set point rate in the last third of pregnancy, but again this is not permanent.

Similarly, he goes on, human set points can change over time.

> There is evidence that we are born with a genetically determined set point, but that is not to say that it cannot be modified. That said, it is resistant to change. If we continue to push our bodies in certain ways, it can be modified. In the laboratory, it is

actually quite hard to make rats get fat. Usually it will only happen when you give them a diet high in fat and sugar, and do not offer them any choice. If you then return them to a normal diet, some return to their previous weight, some don't. More stick at a higher weight if they are heavier for longer, but some are just prone to staying big. Similarly, most people who stay overweight for a long period of time will change their set point.

For this reason, people have started using the phrase 'settling point' instead. Crucially, there is little understanding of the mechanisms that underlie set point or settling point, and although leptin, the hypothalamus and many other hormones clearly play a role, we don't know exactly how the body defines the weight it wants to be. There is also precious little understanding of the factors that make settling points rise, or bring them down again. If you're looking for a single blind spot in the science of why we get fat, this is it: your body has a settling point and it decides what it is without telling you.

If dieting sometimes feels like a battle with your own body, that might be because it is. One thing we do know is that, although settling points are known to move up when people's weight increases, they are far less likely to move downwards after weight loss. Our bodies tend to resist any changes, but they are particularly inclined to resist any loss of fat. Perhaps because of this, studies into calorie-restricted diets show staggeringly high failure rates over the long term (if failure means the long term maintenance of a reduced body weight, which is presumably the goal for the majority of dieters). Some studies have indicated that over 80 per cent of people on calorie-restricted diets will regain all of their lost weight within a five-year period.[5]

To make matters worse, many studies have shown that after the initial period of weight loss, there is a tendency to overshoot to a higher weight when fat is put back on. Many think that this is because a shift in the settling point occurs after the initial calorie restriction, creating often significant increases that would not have occurred otherwise. It is not hard to imagine how this might have developed as an evolutionary response to regular starvation, requiring populations to increase their fat storage in order to see off future threats. A 2007 meta-analysis* by

---

* This is a statistical method of combining the results of several scientific studies to identify any overall patterns. It is supposed to iron out any anomalies and make it easier to find reliable results.

Professor Traci Mann, combining the results of thirty-one weight loss intervention studies, found that between one and two thirds of calorie restricted dieters end up regaining more weight than they lost initially. It concluded that the benefits of trying to lose weight are too small, and the harm too great for it to be recommended.[6]

For many people, dieting can actually make them fatter in the long term. So, not only is it spectacularly ineffective as a healthcare intervention, it seems that it might actually be doing a good deal of harm. This is potentially the most devastating nail in the coffin of the calorie restriction model. But it is also widely ignored, and the study from Traci Mann's laboratory was not considered when the NICE guidelines for obesity treatment were developed.

SHRINKING TESTICLES

When it comes to dieting, there is considerable variability in the data, mostly because of people's inconsiderate desire to live in the real world. This makes dieting, and nutrition generally, a very hard thing to study. Clearly, if people could be made to comply with a highly restrictive diet over the long term, you would see sustained weight loss, something that underlies the regular 'no one fat came out of the concentration camps' argument thrown at those who dare to challenge the 'weight loss is simple' consensus on social media.

Starvation unsurprisingly does induce weight loss, but it can also come at a huge cost. In the legendary 1945 Minnesota Starvation Experiments, conducted by the pioneering nutritionist Ancel Keys, thirty-six subjects were monitored through voluntary starvation. They were placed onto restricted diets that resulted in them losing 25 per cent of their initial body weight, followed by a recovery stage where a variety of strategies were put in place to help them return to normal. The intention was to understand how starvation affected the body, and hopefully lead to the development of effective strategies to help people who had experienced huge weight loss. At the time, this was incredibly important work, with the world reeling from the horrifying hunger and suffering being experienced throughout much of Europe. Most of the volunteers were conscientious objectors, keen to serve their country in a non-violent way, and help alleviate the pains of war. As such, they were a brave and highly motivated group, and as the experiment unfolded, they would need every ounce of that motivation to see them through.

To achieve the desired weight loss, the men's daily dietary intake was restricted to just 1600 calories, a small amount, but interestingly, no less than many well-known weight loss diets recommend today. It was certainly not brutal starvation, and when the experiment was designed, great care was taken to ensure this phase was safe and humane.

Despite this, the effects were devastating. After the experiment had concluded, Keys recommended that it was never repeated, so severe, cruel and life threatening had it become. He became deeply distressed, and was overheard by one of the participants as saying, 'What am I doing to these young men? I had no idea it was going to be this hard.'

He clearly had not predicted how adverse the reaction to this dietary restriction would be. Metabolically, there were profound changes as their bodies tried desperately to fight any loss of weight. Resting metabolic rates dropped by around 40 per cent, body temperatures fell and pulses slowed. The men reported being cold, tired and constantly hungry. They had altered judgement, loss of cognitive function, trouble sleeping, and visual and auditory hallucinations. Their hair thinned, their skin dried out, the size of their testes reduced and they lost any interest in sex.

As well as the high number of physiological changes, many of which are remarkably similar to those experienced by leptin-deficient children, perhaps even more striking were the profound psychological effects. Prolonged dietary restriction led to a condition that Keys termed 'semi-starvation neurosis', resulting in an extraordinary preoccupation with food, as well as widespread depression and numerous other mental health issues. Most subjects suffered from severe emotional problems throughout the experiment, and many reported that they were still greatly affected by it many years later. Participants became socially withdrawn, isolated, self-critical, moody, anxious and depressed. They developed distorted body images, imagining that they were overweight even as their bodies wasted away. They lost ambition, their relationships fell apart and their cultural and social lives narrowed as they became increasingly fixated with food. One of the volunteers was quoted as saying, 'If you went to a movie, you weren't particularly interested in the love scenes, but you noticed every time they ate.'

Many developed strange eating rituals and other obsessive behaviours. Some drank vast quantities of tea or coffee, or constantly chewed gum. A significant number of the volunteers ended up self-harming in some way, and, in the most extreme reaction, one chopped off three of his fingers with an axe.

When the starvation came to an end, and the men entered the regain phase of the experiment, all of them eventually ended up very close to their pre-starvation weights, despite the fact that they were eating in a largely unrestrained way. This was one of the first indications that a biological set point weight might exist in humans. Tellingly, there was no evidence of any of the men reprogramming to a lower weight, which the diet industry would have us believe is a possibility, and the cycling of weight meant that their bodies ended up with a disproportionate amount of fat, as the lean muscle mass they had lost was harder to regain. Most of the volunteers were deeply affected by the experience, and although it would never be allowed today for ethical and safety reasons, the knowledge that it provided lives on. More than anything, it showed the world just how devastating starvation can be, reinforcing a determination to fight hunger above all else. Many of the strides made around the world to defeat famine and alleviate the worst effects of food shortage can be traced back to the profound impact of Keys' work.

Starvation is vicious and consuming, and it is telling how many of the most significant psychological changes seen in the Minnesota experiment mirror those of anorexia sufferers. Starving someone withdraws them from the world, and perhaps for some, this withdrawal becomes a form of escape, a place where the struggles of normal life cease to matter. There is good evidence that a history of dieting is a risk factor for eating disorders, perhaps indicating that finding that particular hiding place depends on having put yourself through food restriction in the past.

Forcing anyone into dietary restriction in order to lose weight is likely to produce similar effects to those seen in the Minnesota experiment, even when their body does not outwardly appear to be lacking in nourishment. Obese people on calorie-restricted diets can reach starvation levels of hormones such as ghrelin, and fire leptin pathways that make the body react to a perceived emergency. This will enact the same changes that are seen in starvation, with exactly the same physiological and psychological effects. In order for people to keep the weight off, these effects will have to continue for years, perhaps even for a lifetime.

BUT STILL WE DIET

There are many different types of diet, many different strategies for encouraging compliance, and a million different factors that might interfere with success and confuse the interpretation of results. But consistently it has been shown that calorie-reduction diets for weight loss, no matter what form they take, do not work for the majority of people. It is a deeply flawed strategy, yet when it comes to the management of obesity, it is the most commonly prescribed intervention. This is not just a matter of poor compliance, which is a problem with many long-term medical treatments. It is a question of hormones. In reducing calories, you are fighting systems that evolved to keep us alive. There is no need to feel guilty when it doesn't work – your body is an expert at finding its way back to your settling point.

But guilty is exactly what society requires people to feel. They have failed in the most basic and fundamental task. They have failed to be a good citizen, to keep themselves safe and well. They will continue to be diseased and ugly, because they could not follow a simple instruction. They could not comply with something so straightforward that it can be said in just four words.

It is the dieter who is left counting the cost of that failure. The lost time, lost energy, the emotional hurt. The lives put on hold until some arbitrary target weight is achieved, magically revealing the mythical thin person trapped within. The more our society equates thinness with moral worth, the more it sees fat as a symbol of degeneracy, the greater the psychological impact of these constant, inevitable rounds of failure. And the more the narrative of simplicity is played, the more the prejudice intensifies.

Do we really think that failing to lose weight through dieting is a sign of moral weakness? That you are flawed if you cannot achieve something the vast majority of people fail at? Are we saying that between 80 and 95 per cent of us are not strong enough? That they don't care? Do we think there has been a fundamental collapse of societal willpower, an epidemic of apathy and laziness?

Or is something else going on? Perhaps something complex and hard to define, beyond the scope of individual will. Maybe the endless articles, blog posts, below-the-line comments and public abuse are doing no good at all. Perhaps stigmatising fat people is actually making things worse.

The pervasiveness of dieting and stigmatisation of fat people doubt-lessly causes psychological harm. But it also appears that our bodies have a particular physiological dislike of the rapid weight loss and regain that is perpetuated by our culture. Those who fail at dieting rarely abandon hope after a single failure, and many will re-engage with dieting time and time again. Often they will return with a desper-ate hope that things will be different this time. Although most will know deep down that another failure awaits, when society defines you as weak-willed and irresponsible if you are not trying, it is hardly surprising that so many get locked into this cycle.

Most of these diets follow a familiar pattern. Initial weight loss will be achieved in the short term, and then regain will occur a few months later. There is a considerable body of evidence that this sort of weight cycling, known as yo-yo dieting, is uniquely harmful to our metabo-lism. It also seems to create an even greater potential for people to overshoot to a higher weight.

The actual harm caused is difficult to assess, not least because neither weight cycling nor yo-yo dieting have a standard definition, making studies of them hard to compare. Popular literature often states that weight cycling is more harmful than maintaining a constant weight, and although there is some doubt about that, it certainly seems to carry an increased risk of hypertension and cardiovascular diseases, kidney disease, cancer, diminished bone density, increased fracture risk, inflammation, harmful changes in body composition (particularly decreased muscle mass and increased abdominal fat accumulation) and increased overall mortality.[7]

DOES IT ALWAYS FAIL?

But hang on a moment, what about the people for whom dieting actu-ally works? For every 80 per cent failure rate, there is 20 per cent of success, and with the prevalence of dieting in our society, that means a lot of people. For every four dieters left feeling like a failure, there is one success story, happily posing for the much craved 'after' picture on Instagram, joyfully holding their old trousers out in front of them to show how far they've come.

To claim that all diets fail is perhaps an exaggeration, as pretty much every calorie restricted programme does produce some short-term weight loss. There is a distinct lack of consistency in the data, with some studies showing a great deal more success than those I have mentioned. There is also evidence that even when diets fail in the long term, a bit of short-term success might be beneficial for people's health. Susan Jebb is the professor of diet and population health at the University of Oxford, and advises the UK government on obesity. She perhaps knows more about obesity research and the success of possible interventions than anyone. She generally has a more positive view of dieting as a strategy for weight loss and health improvement. She told me this:

> Trials tell us that on average, people who try to lose weight succeed. The magnitude of weight loss depends on the type of programme but we have published a series of systematic reviews which show this basic fact very clearly. In fact, even the 'control' group who get minimal intervention are usually about a kilogram lighter a year later. On average people do regain weight, but in trials they are still below baseline four to five years later. While the average weight loss may be less than many people aspire to, we know that it is sufficient to bring health benefits. At a population level, for people who are overweight, weight loss, even small and not sustained in the long term, clearly brings health benefits. Of course more weight loss in more people and for longer is better, but if people don't try to lose weight, we won't achieve anything. We need to support more people to try and choose methods we know from clinical trials to be effective.[8][9]

We shall return to Professor Jebb's work later on. She and her team have done much to help understand obesity, reduce weight stigma, and help people appreciate the level of complexities being dealt with. I am certainly glad I am not responsible for improving the health of the population, especially with so many critics out there. But although such benefits should not be discounted, and for people working to develop public health interventions there is clearly a need for some sort of action, it is hard not to consider the human cost of the huge number of failed diets. When even the most optimistic interpretations of the data show a single kilogram of weight loss after a year, and far less than that over the long term, it is worth questioning whether all that pain is worthwhile.

For any diet, however successful, the question needs to be asked, is the sort of dietary restriction we are recommending likely to provide adequate nutrition? And perhaps more importantly, given the immense societal pressures to comply, is what we are recommending humane? It doesn't seem right that the sort of compulsively restrictive eating behaviours that would result in someone thin being medically sectioned for their own safety are likely to receive nothing but praise if exhibited in somebody who is overweight or obese.[10] Certainly there is increasing awareness that disordered eating can easily exist in overweight people. Kai Hibbard, a former winner of *The Biggest Loser*, was quoted after the show had finished as saying, 'I got to a point where I was only eating about 1000 calories a day and I was working out between five and eight hours a day . . . And my hair started to fall out. I was covered in bruises. I had dark circles under my eyes. Not to get too completely graphic, but my period stopped altogether and I was only sleeping three hours a night.'

In anyone who had not started out as obese, these symptoms would be cause for grave concern, and any competitive television programme encouraging it would be taken off the air. Recent research has high-lighted that eating disorders are often overlooked when weight is seen as a barometer of health, with red flags often being ignored in higher-weight patients. It seems that all too frequently, someone's physical appearance allows us to overlook many severe health problems, including starvation and mental illness.[11]

When people are constantly told that their body size is a disease, a blight, a stain on the good character of the nation, it can be a motivator powerful enough to create disorder, to drive terrible self-harm, and to destroy people's relationship with food and their body. When this is widely endorsed, even praised, then that compulsion becomes even greater. To stray from this path, to take on adequate nutrition, to eat anything other than a starvation diet, is to fail, to be a shameful non-citizen, a drain on us all. We are all complicit in placing these expecta-tions upon fat people.

The profound hunger of a starvation diet can be overcome by will, but how much of an individual's humanity is destroyed in achieving this? Hunger is all consuming, and will stop people from living their lives in a normal way. A world that demands people endure years of suffering, just to fit with bodily norms accepted by our culture, is not one that I find acceptable.

ANY OTHER REASONS NOT TO LIKE DIETING?

Although it is clearly hard for people to lose weight once they are already considered too fat, what story does that failure tell people of normal weight? Is there a warning for us all in the stubborn refusal of fat to melt away by dieting? After all, it is quite possible that many people of normal weight diet just to stay in shape, restricting what they eat to prevent future weight gain, and so never allow their settling point to get too high.

Well, there is pretty good evidence that dieting behaviour might cause harm in these people too, with implications beyond the clinical treatment of weight loss. Large, long-term attitude and behaviour studies like the Growing Up Today Study and Project EAT (Eating and Activity in Teens and young adults) have shown that controlled, restrictive eating and a prevalence of dieting behaviour among normal weight children and teenagers is one of the strongest predictors of weight gain and obesity into adulthood.[12] [13] Behaviours associated with the largest long-term increases in body weight include skipping meals, eating very little, using food substitutes and a history of taking diet pills.

Although much is unknown, there are potential mechanisms that might explain this apparent paradox. As Gareth Leng explains, 'Controlled eaters will have a tendency to disregard signals from the body. They will make themselves constantly hungry, which can lead to leptin resistance. There is a possibility that in dieting, they are undermining their control mechanisms.'

This is a fairly terrifying prospect when you learn that 80 per cent of American girls have been on a diet before their tenth birthday. Seventy-eight per cent of seventeen-year-old American girls are reported as saying they have issues with how their bodies look. Seventy per cent of children in the US between the ages of six and twelve are unhappy with their appearance and would like to slim down. The US has some of the highest levels of obesity anywhere in the world, and yet it is a country utterly obsessed with dieting. Ali Mokdad, the head of the behavioural surveillance branch of the US Centers for Disease Control and Prevention, has been quoted as saying that 75 per cent of Americans are trying to lose or maintain weight at any one time.[14]

Dietary restriction and the control of weight play a huge part in the US national conversation, and are strong drivers of cultural identity in many American people. Americans love to share exactly what they are eating, and particularly which foods they are excluding from their diets this week, usually within thirty seconds of meeting (although to be fair, that might be just when they meet me). Although this sort of obsession may well be a symptom of the high rates of obesity, from what we know about dieting behaviour, it may also be one of the causes. Calorie restriction is known to drive future intake, lowering the resting metabolism, increasing the appetite for energy-dense foods, and perhaps permanently raising the settling point of people's weight. Once that point is raised, people feel fatter, restrict more, and enter a new cycle of loss and regain, overshooting a little further each time. Every time they do, their body will enact powerful hormones to defend an ever increasing settling point. The dieter will get larger and larger, feeling more and more helpless and increasingly miserable. And all that might start with a seven-year-old girl feeling fat and being told she could lose a couple of pounds.

I am not about to claim that dieting causes obesity, because that would be simplistic and reductionist, exactly the sort of thing this book sets out to debunk. But if I have managed to convince you that obesity is mostly caused by an inability of our body to keep its fat stores in balance, rather than an excessive calorie intake, I would suggest it is at least as plausible an explanation for the increases in obesity over the past forty years as 'eating too many pies'. As far as I am aware, pies have not become more delicious, and people have not misplaced their resolve in that time. In a country like the USA, where dietary restriction has become a large part of everyone's life, it might help explain why obesity has reached one of the highest rates in the world.

A CONTROVERSIAL MESSAGE?

If this sounds far-fetched, there is a surprising amount of support for the idea that dieting behaviours in people of 'normal' weight might be a cause of future weight gain. Although some have argued that dieting might just be a predictor of a propensity to eat too much,[15][16] evidence from elite athletes involved in sports such as boxing or weightlifting, where achievement of weight category limits requires cycles of loss and regain, has shown that they have a greater tendency to reach higher weights after they stop training. They not only gain more weight than the general population, they also have a tendency to become fatter than athletes whose sports do not require them to engage in the same sort of weight control.[17] Perhaps more significantly, a large twin study* found that 'normal' weight individuals who engaged in dieting and controlled eating behaviours were more inclined to future weight gain, completely independent of genetic and environmental factors.[18]

There are also a few tantalising studies suggesting potential mechanisms for this, explaining how this sort of overshoot might occur. Much interest has focused on the amount of 'fat free mass' lost when leaner people lose weight, as seen in the Minnesota starvation study, which might upset regulatory pathways.[19] Other potential mechanisms involve the way that our fat cells adapt to weight loss, perhaps changing the way nutrients are transported across cell membranes. Unfortunately, as with the understanding of settling points, these ideas are in their infancy.

Even so, in 2013 this, and other evidence around the potential harms of weight cycling, prompted the 7th Annual Fribourg Obesity Research Conference to conclude that:

> Given the increasing prevalence of dieting to lose weight among those in the healthy normal range of body weight (because of media, family, societal and sports performance pressures) and the emerging evidence that dieting in these population groups is a strong predictor of future weight gain and cardiometabolic risks, one may also add that among strategies to control the obesity epidemic, primary preventive measures should also target these lean 'dieting-prone' population groups.

So a leading group of obesity researchers think that dieting could be making people fat, and we should try to stop people in the normal range from doing it. Others have gone further, with a 2011 review by Linda Bacon and Lucy Aphramor questioning whether overweight adults should be asked to try to lose weight at all.[20]

With the powerful systems at work within the body to prevent long-term weight loss, it should really not come as too much of a surprise that diets fail so often. If this sounds controversial and challenging, perhaps it shouldn't be. A statement from the American Heart Association has

---

* Twin studies allow us to study how genes influence a particular trait or disease. As identical twins share all of their genes, differences between them will be largely due to environmental factors. Obesity is one of the most highly heritable traits ever studied, in the same ballpark as height.

suggested that population-level obesity prevention 'should be approached not as the promotion of widespread dieting, but rather from the perspective of promoting healthful eating and physical activity patterns'. The National Health and Medical Research Council, the main funding body for all Australian medical research, stated in a 2013 report that 'weight regain is common after lifestyle interventions' and that 'weight lost is usually regained by five years of follow up'.[21] Study after study shows the same thing. Dieting does not work for the majority of people, and for many does a great deal of harm.

It is hard not to come to the conclusion that the primary strategy for fighting obesity is deeply flawed, perhaps even exacerbating the problem. It is also extremely tempting to think that our obsession with weight loss by restrictive dieting might just be a contributory factor to the population-level weight gain we are now facing. An obsession with fat, and a tendency to equate thinness with responsible citizenship have created a population that desires weight loss at all costs. The only tools we offer to achieve this are restraint, restriction and control. But what if these tools are doomed to failure? What if they actually make us fat, driving us to gain more and more weight? What if the metabolic harm they cause is exacerbating the problems they were meant to defeat?

Shouting 'eat less, move more' at fat people does not seem to have been working well as an obesity prevention strategy over the past forty years. This has been a remarkably consistent message stretching all the way back to Hippocrates, yet has seen little success. I don't think it is going to start working if our plan is to shout louder and get more abusive. I do not think it has failed because people have not heard the message yet.

Most fat people try dieting. Almost all of them fail. Many of these supposed failures will have great motivation, determination and willpower in other areas of their lives, yet they will not be able to shift the excess weight on their bodies. Some will be experts in motivation, self-determinism, self-control or the workings of the mind. They might be dietitians, nurses, neuroscientists, doctors, psychologists, therapists, business leaders, motivational speakers or writers. They might have been elite athletes in their youth, capable of putting themselves through extraordinary physical exertion and control. They may have fought in wars, showing bravery, dedication and fearlessness. And yet they are fat, and with calorie restriction as the only tool at their disposal, they are unable to become thin. Perhaps the problem is that they have been handed the wrong tools. Or perhaps the real problem is that society is demanding that they become something they are not.

It is perhaps not surprising that there is resistance to these sorts of ideas, because an awful lot has been invested in the calorie-restriction model of weight loss, and clinically we don't have much else in the arsenal. When scientists are asked for an intervention, it is unsettling when they shrug and say there's no simple answer. Any who do are likely to be ignored in a world that promises quick solutions.

Perhaps because most dietary interventions involve professionals working with patients for a relatively short time, maybe six to twelve months, there is a level of clinical bias that leads many to believe instinctively that diets (or at least their particular diets) are more effective than they are. Certainly, with the correct support, people can be physically transformed in the short term, and if clients are not followed up over the three- to five-year period when most regain occurs, it is perhaps not surprising that a rose-tinted view might emerge.

There is also a lot of money to be made from treating the symptom and not the root cause. The constant cycle of failure and desperation is worth $150 billion to the diet industry in the US and Europe combined. If dieting worked, that huge industry would not be sustainable, and although I do not believe in a vast conspiracy, there are plenty of people making fortunes from the false hope that diets provide.

'Eat less, move more' sounds benign and easy to follow, but its logic is vicious, exploitative and controlling. In accepting it, we are accepting that people are fat because they lack agency, which makes them less deserving than the rest of us. 'Eat less, move more' is the foundation on which our prejudices are built, and it gives us an easy culprit for obesity, claiming that interventions only fail because fat patients are flawed, and cannot comply with simple advice or resist the lure of everyday temptations.

I do not intend to lessen the seriousness of the health issues people are facing. But inconveniently, the fact is that what we are doing right now is not working, and is potentially causing a great deal of harm. I don't have a perfect alternative, but neither does anyone else, and we need to stop pretending they do.

WHAT WOULD AN ANGRY CHEF DO?

Perhaps the thing that terrifies me most of all is that, even with everything that I know and the strength of my views on the subject, if I started to gain weight, I am pretty sure I would go on a fucking diet. After a lifetime of thin privilege, I am as afraid of fat as anybody. I am terrified of what it would do to my life, my health, my happiness and my sense of self-worth. I am vain, shallow and fearful of bodily change. I have enormous physical anxieties, and however thin my body might be, I have a lifetime's experience of it rarely doing what I require of it.

I was a gawky, awkward child, who grew into a gawky, awkward adult. My body has a rich and embarrassing history of refusing to respond in the correct way during physical activity, and utterly failing to comply with perceived aesthetic ideals. Like pretty much everybody, I received torrents of abuse over my physical appearance and ungainliness in my youth, something that has continued throughout my life, leaving me profoundly insecure about anything physical. Often I have been angry with my body for not allowing me to be the person I wanted to be, the ideal physical specimen that could have achieved many of the dreams that boys and men hold dear. Although modern society feels progressive and liberated in many ways, it does still place a lot of value on our physique, creating a terrible pressure to conform, something that we all deeply internalise. Although I am now old enough to be comfortable in my physical inadequacy, and have developed many creative ways of telling people to fuck off over the years, as I advance through middle age, fat remains a great fear, and one I hope I do not have to face.

I suspect that almost all of us dislike our bodies in some way, especially as we experience the inevitable decline that comes with age. For me, as my writing has become popular, I have been on television, videos, book covers, magazines, and on stages around the world, sometimes the focus of entire rooms filled with hundreds of people. In midlife I am having to deal with an unprecedented number of images of my dorky, uncoordinated self being thrust out into the world for the first time, and my insecurities scream at me every time one crosses my view. Getting fat into my middle age, a distinct possibility, would be a fucking disaster on so many levels. I am pretty sure that if it happened, I would end up on a diet, despite the knowledge that I would almost certainly fail. So pervasive is the myth, so seductive is the dream, so highly prized is success. I would roll the dice, hoping I was one of the few per cent that wins, not the 50 per cent who end up heavier. I would starve myself, and accept everything that comes with it. That is how afraid I am. Despite writing a book that attempts to blow apart the stigma and shame that surround it, I am utterly terrified, and would do anything to prevent it from happening to me. This is how fucked up things are. The real truth about fat is that we are all afraid.

Right now, without a safe, effective and reliable medical protocol for weight loss, for a huge number of people, things are at crisis level. There is a profound and seemingly uncontrollable epidemic of obesity in our society, creating perhaps the greatest health crisis of the modern age. We are dragging our economy and healthcare systems down into the abyss.

Or at least we are frequently told that is the case. Before we go much further, it is worth looking at exactly how bad the situation is. In the next chapter we shall look at how fat we have become. And once again, although no one can pretend there isn't a problem to confront, we shall see that things are not always quite what they seem.

4

HOW FAT ARE WE?

The obesity epidemic . . . is an extraordinary new phenomenon. It has only emerged over the past forty to fifty years, coinciding with the massive expansion of marketing skills and budgets, principally to promote sales of 'added-value' foods and drinks, and their consumption outside normal meal-times. This is an inescapable truth.
Aseem Malhotra, a medical doctor who sells diet books

It is strange, and perhaps sad, that medical doctors came up with this terminology when they are charged with first doing no harm.
Roxane Gay, *Hunger: A Memoir of (My) Body*

There is a short animation that visualises figures from the US Centers for Disease Control and Prevention (CDC). It demonstrates the rise of obesity from the early 1980s through to 2010, and I first came across it while listening to a talk from Sir Michael Marmot, a renowned professor of epidemiology* and public health. It is a powerful visual demonstration of the inexorable rise of body weight throughout the US, a rise that has been mirrored in many countries around the world. I have subsequently used the graphic a number of times in presentations of my own, and it never fails to shock the audience. It shows how, in 1985, all American states had rates of obesity under 14 per cent, with many considerably below this point. As time progresses, the colours of the states change, with darker colours indicating rates of obesity swiftly rising, first above 15 per cent, then 20 per cent.

* The study of diseases and other health problems across populations.

By 2007, almost all US states have obesity rates higher than 20 per cent and many have turned dark red, indicating levels over 30 per cent. In those states, around one in three people are obese by the time the animation finishes. It appears truly shocking that so much weight gain occurred over such a short period, and it asks profound questions of anyone who is interested in food, health and the way our genes interact with our environment. Clearly, in the years covered, something dramatic has occurred. A wave of obesity has swamped the US and the rest of the world, leaving nowhere untouched. A devastating apocalypse. An epidemic.

Michael Marmot's talk also highlighted the vast inequalities between rich and poor when it comes to obesity, claiming that the only way to tackle these issues was to create a more equal society. At the time, I thought this quite absurd. Surely if we could get a better grip on the reasons why poorer people are gaining weight, we could find ways to influence behaviour in order to improve their health. Socially deprived groups have not always been fat. In fact, until recent times, quite the opposite was true.

But setting this aside for a moment, obesity's rise has been quite extraordinary. In the UK, 2012 figures showed that roughly a quarter of all adults are classified as obese, with a further 42 per cent of men and 32 per cent of women falling into the overweight category. Obesity is strongly related to socio-economic group, something grimly detailed in the 2010 Marmot Review into health inequalities. Across the country, the prevalence of obesity in the most deprived population groups is nearly double that of the wealthiest. Figures are particularly stark when it comes to children, where rates of obesity for pupils in the last year of primary school rise from below 12 per cent for the most well-off children, up to 25 per cent for the most deprived. Obesity has also been shown to have strong links to educational attainment. In the UK, 20 per cent of people with a degree-level education are obese, a figure that rises to over 30 per cent for people with no qualifications.

In 2017, the Organisation for Economic Co-operation and Development (OECD) calculated that the UK had the highest rates of obesity in Western Europe. Tellingly, Britain also stood out from the rest of Europe in another poll, conducted around the same time. It appears that the UK was more inclined to blame individuals for their obesity, with 41 per cent agreeing with the statement that people should be 'blamed and faulted for their condition'. (Perhaps we haven't been abusive enough yet.)

A few things are certain. Obesity has risen hugely and in a short space of time, with a marked increase occurring around the late 1970s and early 1980s.[1] Rates appeared steady throughout much of the twentieth century, then, in the twenty years between 1980 and 2000, they doubled in the US, with the UK not far behind.

Once the preserve of a privileged few countries, obesity is now a worldwide health issue. While the rises in the US have started to level, some of the most dramatic recent increases have been seen in Mexico, China and Thailand.[2] [3] It seems to follow a familiar pattern, initially affecting the higher socio-economic groups, before shifting to the poorer demographic as time progresses. Rises tend to be correlated with a Westernisation of economies, and the introduction of Americanised values and produce. Due to the poor quality of many people's diets, and the general lack of financial resources, often obesity and undernourishment end up existing side by side within the same communities. Sometimes, the two are seen together within a single family. In modern day Brazil, it is not uncommon to see an obese father living under the same roof as an undernourished child.

Theories abound as to what happened, and many proposals for wide-scale public health interventions have been based on intuitions and hunches. Most ideas centre on changes in food production, marketing, and dietary guidelines, but also on declines in physical activity, suggesting an environment that encourages people to 'eat more and move less'.

In recent years, some of the increases have abated, with levels stabilising across wealthier countries, but failing to fall in any significant way. Where successes have been achieved, often these occur only within the most privileged demographic groups, leaving rates among the disadvantaged static or worsening. Obesity, once a problem of affluence, has become an issue of poverty, and the divide is only increasing. More and more, as interventions fail to reach the most deprived members of society, weight becomes a visual signifier of social status, and links between thinness and privilege grow ever stronger. Weight is a class issue in a way that it has never been before, and in recent years this has vastly changed our perceptions of what it means to be fat.

BMI

The inexorable rise of obesity across such large swathes of the population is almost beyond comprehension, indicating a dramatic shift in our lifestyle and behaviour. But before drawing conclusions, it is worth stopping for a moment to think about what these figures actually mean. How exactly have so many people become obese?

Key to understanding this rise in obesity is the method by which it is determined. Obesity occurs when someone crosses a threshold of BMI, a number that is calculated by dividing your weight in kilograms by the square of your height in metres. For instance, if you happen to be 2 metres tall and weigh exactly 100 kilograms, you would divide your weight by four to get a BMI of 25. The history of this calculation is interesting, because although it has been around for some time, it was never originally intended to be used as it is now.

In the 1830s, the Belgian polymath Adolphe Quetelet engaged himself in a project to categorise the 'normal man' and noticed, perhaps unsurprisingly, that people's weight did not vary in direct proportion to their height. Instead, he found, weight differences were more closely correlated to their height squared, and he created what became known as the 'Quetelet Index'. His work was not an attempt to study fat, and the index only represented a rough statistical observation, but it was interesting because it indicated that as people's height changed, so did their body shape. For this reason, the index was only ever an approximation, highly unlikely to apply to people of all shapes and sizes. In Quetelet's studies, it was not applied to women, children, the elderly or people of different ethnicities, and was never used as a measurement of bodily fat. Quetelet never had any intention for his index to be used in a medical capacity, and wouldn't have known that it would be, since it was pretty much ignored for the next 150 years.

It was not until 1972 that a potential new application of the index was first realised. Ancel Keys, the same nutritionist who conducted the starvation experiments discussed in the last chapter, realised that in a study of 7,400 men, measured body fat percentages could be closely predicted using Quetelet's formula. It was Keys who coined the term 'Body Mass Index' and helped to popularise it as a commonly used marker of health. It was never going to be perfect, but it did appear to be a reasonably good predictor of levels of stored body fat when investigating large groups of people. More recent work has indicated that it accounts for about 60–75 per cent of the variation in body fat, which for population studies is not that bad.[4] But as a tool to diagnose the health status of individuals, that margin of error isn't going to work.

Although Keys recognised that BMI was a potentially useful tool for assessing prevalence of fatness in large groups of people, the paper in which he first described it contained a number of reservations about its use in categorising individuals.[5] He considered it a blunt statistical tool, something that takes no account of important factors such as age and sex, both of which greatly influence the relationship between BMI and body fat. The paper did not cover a number of other potential confounding factors* highly likely to influence the relationship, including ethnic group, body shape, bone density, muscle mass and the differences seen in children as they develop. Perhaps because of an awareness of these issues, the concluding statement broadly criticises the use of terms such as 'ideal' or 'desirable' when describing any measure of body weight, saying that 'it is scientifically indefensible to include a value judgement in that description'.

In 1979, Keys went further, roundly attacking measures being used to divide people into weight categories, saying that such things were 'arm-chair concoctions starting with questionable assumptions and ending with three sets of standards for "body frames" which were never measured or even properly defined'.

After Keys' paper, BMI was rapidly adopted, particularly by insurance companies, which were keen to find simple statistical indicators of health to use in their calculations. BMI figures were cheap to produce, easy to calculate, and seemed to be a half-decent predictor of how much fat people were carrying around with them. The simplicity of the calculation, usually requiring no new measurements to be taken, meant that people could delve into ancient data sets to assess changes in body fat through time, looking for correlations and new understandings of risk.

Despite Keys' warnings, BMI also came to be widely accepted by the medical community as a marker of individual health. Today it is

---

* A confounding factor is a variable that influences two things separately, so creating a correlation between them. For instance, deaths by drowning and consumption of ice cream are closely correlated, which might lead you to conclude that ice cream makes people drown. But here, the confounding factor is hot weather, which leads to an increase in people swimming, and also an increase in people eating ice cream. To be honest, if you had to look this up I am disappointed that you did not read my last book, as confounding factors was Chapter 1. It had a really nice analogy about hares, lapwings and Easter eggs, and took me ages.

routinely measured by health professionals, with care guidelines around the world often broadly based on BMI classifications. The NHS website contains a BMI calculator, encouraging people to diagnose themselves, and most of us will be aware of where we sit in the current system.

BMI is used to classify people according to their supposed body fatness, with thresholds for underweight, normal (healthy) weight, overweight and obese. According to the World Health Organization (WHO), people should be classified according to their BMI as follows:

BMI	Classification
Less than 18.5	Underweight
Between 18.5 and 25	'Normal' or 'Healthy' Weight
Between 25 and 30	Overweight
Between 30 and 35	Class I Obese
Between 35 and 40	Class II Obese
Over 40	Class III Obese

Within these categorisations there is much controversy, and some significant differences among countries. Many feel the underweight threshold of 18.5 is too low for certain racial groups, with 20 being a more accurate cut off, but that would leave a good number of healthy Asian women with an underweight diagnosis. In some countries, including Hong Kong and Singapore, the overweight category is defined as being anything above 23, mostly to reflect evidence suggesting an increased risk of diabetes and heart disease in those populations at lower weights.

In children, the use of BMI is further complicated, because growing and developing bodies change rapidly in shape and size. A BMI of 20 in a five-year-old boy would probably indicate a good deal of excess fat, whereas the same BMI aged fifteen would indicate a child on the lean side. For this reason, there are no fixed thresholds, and children's BMI measurements are compared against standard population tables dependent upon their age and sex. This comparison produces a so-called z score, which can be used to assess whether that child is normal weight, overweight or obese. The tables used generally come from historical data, and their relevance is dependent on all children developing and growing in the same way, something that is clearly not the case. The tables used vary among countries and are often based on data that is many years old.

Childhood BMI is actually thought to be a relatively weak predictor of body fat percentage, as young children tend to have a greater proportion of fat-free mass making up their weight. And the often touted relationship to negative health outcomes is also highly questionable. A 2012 systematic review concluded that there was little connection between childhood BMI and the development of metabolic diseases into adulthood, with the greatest risks affecting lean children who go on to become obese adults.[6]

In the UK, Public Health England uses data from 1990 and considers a child whose BMI would have been in the top 5 per cent for that year to be obese. This assumes that 5 per cent of children were obese in 1990, something for which there is no evidence and was almost certainly not the case. As a result, many children are currently labelled overweight or obese right up until their sixteenth birthday, but when they pass that threshold and are judged instead by the adult criteria, they are suddenly considered of normal weight. This might explain the shocking media reports that claim many parents can't see their own children are obese – perhaps, by any sensible measure, they are not.

It could be said that when it comes to assessing changes within populations, as long as there is a degree of consistency, the particular table used is not all that relevant. But when BMI becomes a tool to assess the health of individuals, we run into problems. As part of the UK's National Child Measuring Programme, all children in the first and final year of primary school have their height and weight measured and their BMI calculated using Public Health England's standard 1990 tables. This no doubt provides useful information about population-level changes over time. But the information is then individualised and sent to parents, telling them the height, weight and classification of their child. The flawed method of calculation, and the vastly differing rates of growth across the population, mean that many children have been classified as overweight when a simple visual check would have confirmed they were nothing of the sort.[7] [8]

There have also been indications that in some cases the use of BMI to assess children might actually under-report rates of obesity,[9] and suggestions that other measures of body size should be used in combination to provide a more accurate picture. At the very least, if it is really necessary to make children and parents aware of weight concerns, they at least deserve a medical professional taking a cursory glance at the child in question. Faceless letters based on raw and unreliable data are hardly an effective way to help people. BMI is already flawed when it comes to assessing the health of adults. But in children, its use is absurd.

EVERYONE LIKES A NICE ROUND NUMBER

Obviously, within the nuances of age, race gender and body diversity, it is difficult to provide accurate BMI cut offs that suit everyone, which is perhaps why such a project has proved so divisive and troubling. The value judgements that so concerned Ancel Keys are alive and well, created by the WHO and embraced by healthcare professionals around the globe.

The adult cut off for the overweight category of 25 is perhaps the most controversial. Until 1997, anyone with a BMI below 27 was considered of normal weight, but a WHO report, partly based on pre-war data from a US insurance company, recommended that it should be moved down to 25. This reflected a perceived increase in mortality risk, but was also a nice round figure that was easier to remember. The WHO report that made this change was called 'Preventing and Managing the Global Epidemic'.

The data on which this decision was based was deeply questionable, and has been criticised frequently over the years, but the 25 cut off has remained firmly in place. Millions of people suddenly woke up to find that they had a health problem. Around thirty million sick people had been created overnight, all requiring some sort of medical intervention. Quite a few of them will have decided on a restrictive diet to 'cure' themselves of their new illness, which may well have been the only real cause of harm.

It also seems remarkably unlikely that the cut offs at 30, 40 and 45 were set that way for anything more than the convenience of being round numbers, and it has often been reported that this was done to make them easier for the public and medical professionals to remember. Considering this, people moving from one category to another is little more than an arbitrary shift, perhaps only indicating a weight gain of a pound or two. But in terms of how those individuals are viewed by the world, such a change can have dramatic consequences.

THE OBESITY 'EPIDEMIC'

As well as making millions of people overweight (at least by definition), the 1997 WHO report classified obesity as an 'epidemic' for the first time. It also classified anyone with a BMI over 25 as being 'pre-obese', indicating that once you crossed the new lower threshold, you were destined to slip into more damaging pathologies, despite presenting no evidence that this was the case. Obviously, everyone who is obese has been through at least a brief stage of being overweight, but it does not logically follow that everyone who is overweight is at risk of becoming obese. That is a bit like saying any child who is five feet tall will go on to reach seven feet six.

The report created a damning picture of an unstoppable epidemic spreading around the world, likely to have devastating consequences for global health systems and economies. But crucial in defining future discourse on the subject was the use of the term 'epidemic', something that was certain to cause alarm.

A brand new 'epidemic' proved to be a perfect way to provoke mass social panic, instantly resulting in calls for dramatic action and interventions. This vision of a sweeping and unstoppable disaster, a contagion spreading from country to country, was irresistible for the media. News items began to feature terrifying dehumanised images of fat bodies, frequently with pixelated faces, or cut from the head down. They would lumber aimlessly around, stuffing burgers into their hidden faces, in images similar to popular visions of a zombie apocalypse. 'The fat' were coming, and they were going to drag down the world.

Because epidemics are often contagious, they carry with them an expectation that everyone will act for the greater good. We all become obliged to take steps to protect ourselves, our families and our communities, even when this means great personal sacrifice that we would otherwise find unacceptable. Whatever it takes to win the battle, we are willing to entertain desperate actions from the authorities. We dutifully comply with everything that is asked. The term carries with it an implication of contamination and spread, an idea that there are 'others' out there, and contact is harmful. The 'epidemic' was no doubt designed to inspire action, yet it also bred associations of uncleanliness and danger, played out almost daily in the news media.

An epidemic also suggests the search for a vaccination or a cure. Disease epidemics have a history of successful public health action, where causal agents are cleverly discovered and ingenious interventions devised. In 1854, when John Snow removed the handle from a pump in London's Broad Street, he managed to identify that a contaminated water supply was the causal agent for a devastating cholera epidemic. This famously led to successful public health interventions that saved millions of lives, and allowed the safe urbanisation of cities worldwide. With obesity now defined as an epidemic, the race was on to find the reason why it was spreading uncontrollably around the world. Halting the fat zombie apocalypse became science's primary mission.

Moral panic and outrage is potentially very motivating, and throughout history has been the force behind much positive change. But when there is no single cause that everyone can agree upon, panic amounts to little more than running round in excited circles. At its best, it is a waste of time and energy. At its worst, it creates division and harm. Before John Snow identified water supplies as the source of cholera, many thought it was obvious that the disease was caused by the stench of human and animal waste left out on the streets in the poorer parts of town. As a result, tonnes of waste were shovelled into the Thames to clear the stench. In an effort to 'do something', the shit was piled right back into the water supply, doubtlessly killing thousands more.

PROFITING FROM OUTRAGE

Whenever there is a society-level problem, the right wing finds a way to blame an individual's moral fibre, and the left wing finds a way to pin it on big nasty corporations.

With obesity now an epidemic, neoliberals* finally had an excuse to blame a global catastrophe on the behaviour of irresponsible individuals, pitting a lean, ambitious elite against a feckless group of fatties. It encouraged those of normal weight to look down upon the fat, to see them as 'others', not smart or ambitious enough to lower their BMIs. This absolved the political right of any responsibility to address

* Neoliberalism is a political ideology that favours free-market economics, and is associated with a focus on individual responsibility, blaming people's circumstances on their character, rather than on the blind luck that underlies most wealth and privilege. This is clearly mad, and perhaps creates the sense of entitlement that leads obnoxious groups of Old Etonians to trash restaurants and stick their cocks in the mouths of dead pigs.

divisions and inequality, and also conveniently allowed them to blame a crumbling state health service on the selfishness of fat people, rather than on years of underfunding and neglect.

On the other side of the political spectrum, things were little better. The left framed the epidemic as a problem of corporate irresponsibility, thus demanding regulation of the evil conglomerates that were themselves growing ever larger. They called for taxation to curb sales of the brands they despised so much, with no regard for how these taxes would hit the most vulnerable in society. Some called for a complete centralisation of the food supply, others for the prohibition of offensive foodstuffs. The left lauded folksy initiatives that carried vague notions of naturalness and local provenance. They demanded a return to the golden age of war rationing, conveniently ignoring the devastating rates of undernutrition that millions were forced to endure.[10] For the left, the obesity epidemic was a demonstration of the abject failure of a free market economy, a grim reminder that corporations could not be trusted, and populations needed to be regulated and controlled. There was now a global epidemic to fight, which meant taking emergency measures.

Drug companies saw rich profits to be made. An effective pharmaceutical intervention for obesity would be a blockbuster of epic proportions, of immense value to whoever got there first. There was rightful criticism of the fact that a number of drug companies were funding the International Obesity Task Force when it helped to produce the 1997 WHO report, as the creation of a tasty new epidemic was clearly in their interest. Share-price rises created by even the slightest hint of an effective pharmaceutical treatment could easily create millions of dollars in value.

Journalists and media commenters embraced the epidemic too. They now had a compelling new story, a dramatic battle against a terrifying, relentless foe. There would be heroes, villains, conspiracies, breakthroughs and disasters. It could be spun to suit whatever political persuasion you wanted to support. There was a raft of military metaphors connected to the fight, including the commonly used 'ticking time bomb', 'battling', 'fighting' and 'waging war'. When there hadn't been a photogenic natural disaster in some time, fat became the 'rising tide' that threatened to consume us all. It was a disaster movie made real, but played out slowly, providing years' worth of valuable column inches.

Food companies saw huge opportunities, as did retailers. Suddenly diet and weight loss brands were frontline heroes, rather than the joyless section no one reads in lifestyle magazines. Commercial weight loss clubs, once hidden away at the back of leisure centres on a Tuesday evening, were now prescribed by healthcare professionals and proudly advertised in national newspapers. Diet products flew off the shelves, and 'health' became the biggest trend in food. It did not matter that the definition of what constituted 'healthy' was poorly defined. A prominent food marketer, who worked on baked beans at the time, once told me that it did not matter what you actually claimed, just so long as you had plenty of ticks on the front of every pack. Low in fat. Tick. Low in sugar. Tick. Free from dairy. Tick. Tick. Tick.

The diet food industry did not care that selling foods designed to help people lose weight was bizarre and oxymoronic. All that mattered was that there was a new category, and in a troubled food sector, 'health' was the only area experiencing any growth. Brands with a health halo flourished and grew. Weight loss brands that were once tired and dated exploded into the mainstream. There was money to be made in exploiting people's fears, and the food industry seized every opportunity with both hands.

The obesity epidemic brought out the worst in people. Soon the medics and surgeons appeared, offering us dramatic solutions to save us from ourselves. They would cut and dice people's stomachs in dangerous experimental surgeries. Then the conspiracy theorists crawled out of the woodwork, with endless YouTube videos claiming that dark capitalist forces were keeping us all fat. The diet sellers started to cash in, merchants of certainty who claimed that everything you know about food is wrong, then helpfully offered complete solutions based on their own special brand of research. Cut fat. Cut sugar. Eat clean. Eat paleo. Eat fat. Eat protein. Cut carbs. Count calories. Don't count calories. Work out. Don't work out. Buy my video. Read my website. Buy my magazine. Buy my products. Buy my book. Buy my book. Buy my book.

Because so many people had an interest in the creation of an obesity epidemic, few ever stopped to criticise the language or the rhetoric. It created valuable companies, driving growth and profits for a number of struggling industries. To this day, thousands of people are employed in service of the obesity epidemic, and many would suffer a great hit to their income should it come to an end.

The ones who suffered were fat people. With no magic bullet ever likely to be identified, they were left in the wilderness, alternately told that they were weak-willed or easily manipulated. Most of all, everyone who accepted the epidemic narrative accepted that fat was a disease, something to be quarantined, perhaps even something contagious. It was a zombie apocalypse, and there was no compassion for collateral damage. This was despite increasing evidence that the sort of shame and vitriol being levelled at fat people was actually making things worse.[11]

THE STING IN THE TAIL

Clearly, the WHO defining something as an epidemic means it is probably an area of serious concern, but obesity is neither a disease nor a behaviour. It is quickly becoming clear that obesity will never have a single cure or strategy that will work for everybody.

It is also worth considering exactly what happened to create this epidemic. There is no doubt that the use of BMI thresholds to define obesity was a big factor. Figures showing a doubling of obesity in twenty years often accounted for a large number of people moving from a BMI of 29 to 30, which might have represented a shift of only a few pounds over several years. A 1998 US population study by the epidemiologist Katherine Flegal showed that although obesity had doubled, the majority of people had increased in weight by around 3–5 kilograms in a generation.[12] [13] This was significant, but hardly the zombie apocalypse represented on the news. What had changed, quite dramatically, was a substantial reduction in the number of underweight people, universally regarded as a very good thing for population health. And troublingly, there was also a large increase in the number of people who were severely overweight.

When the BMIs of a population are represented graphically, they will generally produce a bell-shaped curve in much the same way as other continuous variables such as height or IQ. If BMI runs in numerical order along the bottom axis, the graph will show that small numbers of people on the left are underweight, with BMIs of 18.5 or below. The curve will rise up in the middle as BMI increases, showing that the majority of people have a BMI in the middle range. As we move to the right, the number of people tails off, forming the right side of a roughly symmetrical bell-shaped curve. The strange thing about the increases in BMI since the 1980s is that the curve has not just shifted a few points to the right, it has actually changed shape. The number of people on the far left of the curve has decreased, and the number of people on the right has increased, meaning that even though average weights have not moved a great deal, the far right tail of the curve has been growing considerably.

A 2003 report showed the extent of that movement in the US, with a large increase in the number of people represented in that right tail.[14] In that period, the number of people with a BMI of over 40 had increased four times, rising to 2 per cent of the population. The number with a BMI over 50 had increased five times, reaching 0.5 per cent. Although these numbers are still relatively small, combined with a reduction in underweight people, they have significantly increased the average weight. In the UK, the figures show a similar trend. Between 1993 and 2014, cases of severe obesity, defined as people with a BMI over 40, more than doubled, despite relatively small shifts in average body weight. Although hugely concerning for this small minority, in terms of the health outcomes for most people it is largely irrelevant, because the vast majority have not seen that much change.

It is also worth remembering that much of the population data on obesity is not corrected for age, and as many of the countries affected have increasingly ageing populations, a rise in average body weight of just a few pounds becomes even less surprising. It is generally expected that most people will gain a little weight as the years progress, and there is very little evidence that such moderate increases are in any way harmful.

Professor Jeffrey Friedman is a medical geneticist at Rockefeller University who has added more to the field of obesity research than perhaps anyone alive. His research team identified the ob/ob gene in mice and humans, and so our understanding of the hormone leptin. He even gave it its name, after the Greek word *leptos*, meaning thin. In interviews, he has vividly described how the use of BMI thresholds can be used to manipulate statistics and create the illusion of an epidemic:

Imagine that forty years ago the average IQ was 100 and there was a bell-shaped curve. Imagine now that our educational system improves and the bell-shaped curve shifts a little and the average IQ is now 105. With that you could imagine that the

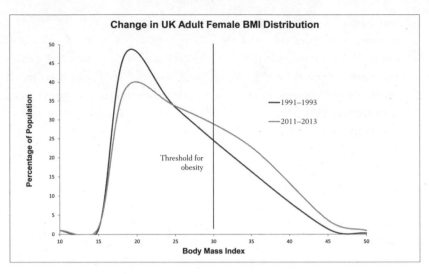

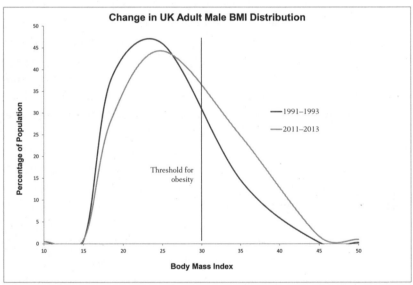

Both sets of graphs show how average BMI has shifted to the right slightly, but the graphs have skewed and changed in shape. Note how this puts many more people into the obese category.

Sources: UK graphs from Health Survey for England 1991–1993 and 2011–2013.

number of people who have an IQ greater than 140, so-called geniuses, might have doubled. Now, is it more useful to think about how our education is doing by saying, 'average IQ increased 5 points' or 'number of geniuses is doubled'? I think probably both are of interest but the former seems to me more informative. OK. So how does that analogize to weight? Over the time period that you've heard that the obesity rates have 'doubled' or gone up by 70 per cent, the average weight gain is 7 to 10 pounds . . . think about the fact that 7 to 10 pounds is almost nothing compared to the hundreds of pounds of difference in weight that you might see in any two people walking around the street today, both of whom essentially have unlimited access to calories.[15]

Strangely, a story that 'people have gained a few pounds in twenty years as they've got older' is not likely to sell newspapers. This is why average weights and BMIs are rarely discussed in the debate on obesity, because then the problem seems potentially less dramatic, perhaps more of a curiosity than an epidemic. But these are the figures that truly matter to people, the ones by which we will compare the epidemic to our own lives and experiences. A few pounds over twenty years might produce an interested shrug, but frame the same numbers differently and an epidemic is born.

THE RIGHTWARD SKEW

This right skew and the changing shape of the obesity bell curve is one of the most important aspects of the rises in population body weight, because it strongly indicates that instead of a worldwide epidemic that threatens us all, obesity has mostly affected a small number of vulnerable individuals. Average heights have increased appreciably over the same period, but with them the bell curve remains much the same shape, just moving slowly rightwards. This indicates that unlike with height, the changes in weight depend on different people reacting to the environment in fundamentally different ways.

Understanding these differences is key to understanding how the problems should be addressed. It seems highly likely that the people reaching BMIs of 40 or 45 might be doing so for different reasons than the majority, most of whom have only experienced limited gains. There will never be a one size fits all solution, yet that is what we crave, and that is what is constantly being searched for. As a result, people genuinely in need of help are cast aside as wide reaching interventions are designed on the premise that they might shave tiny amounts of weight from the whole population. In trying to 'cure' everyone of these few extra pounds, we are actually helping no one.

Changes in height and weight, 1959–2010

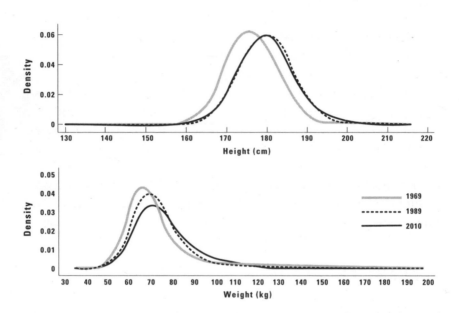

Note how in this data (German, from 1989–2010), the bell curve for height simply shifts to the right, whereas the bell curves for BMI and weight change in shape, skewing to the right and developing a fatter tail.

Source: Lehmann et al., 'Temporal trends; regional variation and socioeconomic differences in height, BMI and body proportions among German conscripts, 1956–2010', *Public Health Nutrition*, February 2017, 20 (3), pp. 391–403, https://doi.org/10.1017/S1368980016002408.

OH, AND ANOTHER THING . . .

There is another trick commonly used to overplay the scale of the epidemic. In isolation, the obesity numbers do not appear quite as shocking as they do if you use 'overweight and obese' as a catch-all term. Because of the large numbers who have crossed the BMI 25 threshold in recent years, figures of 60 to 70 per cent are now common, and we are left horrified and appalled that most of us are now overweight, rather than focusing on the small numbers of very fat people in need of help. Being a little bit overweight, as we shall see in the next chapter, is of questionable harm and is not necessarily a gateway to future obesity, so to lump the terms together doesn't make much sense. But if you take away the ubiquity of the epidemic, obesity loses much of its political power, and there are many people with an interest in that not happening.

Cases of severe obesity are of significant concern for many reasons, and no one can deny the medical harms. But the term 'epidemic' is unhelpful at best, frequently leveraged for political gain, with the potential to hurt a great many people. It has added to the demonisation of fat people on a worldwide scale. But perhaps even worse, it has left us looking for population-level solutions to a slight rise in the average of a crude measure. And contrary to what many will have you believe, there is also some evidence that a rise like this should not have come as a surprise.

A RECENT PROBLEM?

The search for a magic bullet has often focused on the widely accepted fact that rapid weight increases started in the early 1980s, and theories abound as to what events might have promoted such a change. Leaving aside for now the large and concerning increases of the most serious cases of obesity, the wider population-level changes were not as unpredictable as some people make out. Studies of BMI from historical data have shown that it might actually be little more than a statistical anomaly, which has serious implications for anyone trying to create policies to reverse it.

Most obesity statistics are snapshots of the whole population at a particular point in time, but if there are any large demographic changes, the data can provide a confusing picture. This is especially the case over the last half of the twentieth century, when many dramatic shifts in population occurred due to mass migrations, increasing life spans and the infamous baby boom generation.

In order to counter some of these effects, people born around the same time are sometimes studied as a group or cohort, in something known as a 'birth cohort analysis'. When this is done with the obesity statistics, a slightly altered picture emerges of how body weight has changed over time.[16] In many ways this sort of analysis makes sense, because different generations are likely to respond differently to their environment, having had different upbringings and life experiences. It also takes into account the fact that most excess weight is gained slowly over a lifetime, rather than being fixed at any particular age, removing some of the confounding effects caused by increasing numbers of elderly people.

Studying the figures in this way shows that, having remained low for many years throughout the nineteenth century, BMIs in industrialised nations started to increase from around 1920. It is perhaps no coincidence that this occurred at a time when food supplies became more secure and some of the most savage hardships people faced were addressed. In the US, between 1900 and 1960, BMIs increased around 0.5 units per decade. From 1960 onwards, they increased steadily by around 0.7 units per decade, right up to the end of the century. When analysed in this way, there is little evidence of a rapid growth spurt around 1980 – instead, there's just a steady, relentless progression from the 1920s onwards, slowly increasing as people become more affluent. Throughout the twentieth century, eighteen-year-old men increased in weight by an average of 13 kilograms (40 pounds), but more than half of this increase had already occurred before the start of the Second World War. These sorts of modest, steady changes perhaps just reflect the way that human populations alter when food is no longer scarce. It might be that the obesity epidemic is not some terrifying artefact of modernisation, but something that happens when the threat of starvation is taken away.

It is tempting to draw conclusions from birth cohort data, but it does have limitations. In truth, what it really provides is a lower boundary for when changes might have occurred, failing to account for environmental factors that affect people across the generations at any one point in time. It is, however, a strong indication that things are not quite as simple as they seem, especially for anyone claiming that weight gain was a mysteriously 1980s phenomenon. The truth probably lies somewhere in between, with steady increases throughout the twentieth century and a slight acceleration towards the end.

BMI in the Twentieth Century

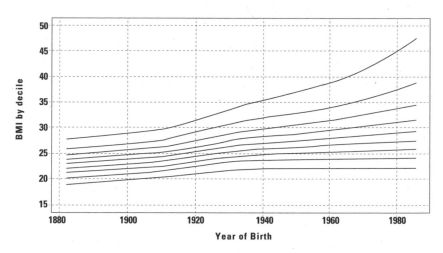

Year of Birth

This graph shows the slow and steady rate of change in BMI throughout the twentieth century by birth cohort at roughly 1 BMI point every two decades, but also how the highest decile has grown far more rapidly than the lower ones (i.e. the fattest people are getting fatter). Different people have a dramatically different reaction to increased food availability and societal changes.

Source: Komlos and Brabec (2010), 'The trend of BMI values of US adults by deciles, birth cohorts 1882–1986 stratified by gender and ethnicity', *Economics and Human Biology*, 9 (2011), pp. 234–50.

Although visual demonstrations of obesity rates are commonplace in the media, graphs of how the average BMI has changed over the decades are rarely seen, perhaps because they reveal a less dramatic picture. Whichever way they are studied, increases in BMI have been fairly steady, something that has produced gradual rightward movement in the familiar bell-shaped curve. As a result of this, in the late 1970s, the steep right-hand side of the curve started to cross the 30 threshold, meaning that obesity numbers were always likely to increase rapidly around this time.

Even Professor Katherine Flegal, the epidemiologist who famously identified the dramatic increases in US obesity prevalence while working as senior scientist at the National Center for Health Statistics, had this to say in a recent interview:

Although these phenomena are often referred to as the obesity 'epidemic', the word may be misleading. An epidemic is a number greater than expected, but what would be the expected prevalence of obesity? The prevalence in 1960, which was about 15 per cent? Obesity might more properly be regarded as endemic, not epidemic. Data dating back to the [American] Civil War, though limited, suggests that weight has been increasing fairly steadily ever since and increasing at a slower rate now than it did in the last half of the 19th century. A broader perspective then would suggest that the current trends in obesity are a further manifestation of this longer-term trend and not a sudden outbreak of a disease.[17]

BMI Values and Weight of Eighteen-year-old American Men

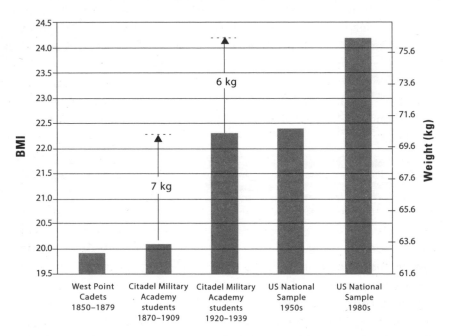

The largest changes in BMI occurred in the period between the First and Second World Wars. Increases have been going on for well over 100 years.

Komlos and Brabec, 'The evolution of BMI values of US adults: 1882–1986', VOX, 31 August 2010, https://voxeu.org/article/100-years-us-obesity.

In truth, when anyone asks the question why obesity started to rise around the 1980s, the correct answer is: because the BMI cut off for obesity is 30. If it had been 32, similar increases would have occurred a few decades later. If it had been 28, people would have started becoming obese many years before. And remember, one of the main reasons why the obesity threshold is set at 30 is because it's a nice round number that's easy to recall. Drawing conclusions about changes in the 1980s making us fat is a dangerous game.

WHAT HAS REALLY CHANGED?

Despite all this, the fact remains that people are getting fatter, and although the increases are not nearly as dramatic and unexpected as have been made out, this does require explanation. We have already discussed the rise of diet culture, but it is also worth considering some other factors that might have played a part in any late-twentieth century accelerations in BMI, many of which we might not want to reverse. We shall look at some more of these in Part II, but for now it is worth mentioning briefly how the decline of smoking might have played a significant role.

The appetite control effects of tobacco were well known to a generation of 1990s supermodels, and will also be familiar to anyone who has ever smoked. The work of a chef frequently requires you to go without meals and during my younger years, like many in the profession, I often used a cigarette break as a convenient, if inadvisable, meal replacement. Rates of smoking in the UK and US have been declining steadily for the past fifty years, and a 1995 study,[18] again by Katherine Flegal's team, estimated that around a quarter of the increase in the prevalence of overweight men could be accounted for by them stopping smoking in that period. It is thought by many that wide-scale twentieth-century increases in obesity were repressed by high rates of smoking up until the 1960s, with their true extent being revealed as more and more people decided to quit.[19] This is a good answer to the claim that obesity is caused by feckless individuals who don't care about their health.

It is unlikely that encouraging an increase in cigarette smoking will ever be adopted as a public health strategy. Neither will a return to the Victorian-era poverty that kept BMIs down throughout the nineteenth century (although one or two campaigners I have come across might go for it). But if we view weight as something that has seen a slow rise since the 1920s, obesity perhaps becomes a different issue, to be perceived in a different way. Instead of demonstrating the failings of modern society, it represents the ways in which absolute poverty has been addressed. It shows how mass starvation events are now less common than they have ever been, and how our industrialised food system has triumphed over hunger, the greatest and most dehumanising threat facing us. It shows us how our ingenuity has won the fight against many communicable diseases, and how the health and longevity of populations have increased dramatically in such a short time. It shows how our food supply is safer and more reliable than ever, leading to dramatic reductions in gastrointestinal disorders, poisonings and contamination events. It shows how we no longer accept hunger as an acceptable risk of living, and demand universal access to adequate nutrition as the most profound right of all humanity.

Generations of people throughout the twentieth century fought against hunger and dedicated their lives to keeping us fed. For those of us who have never experienced starvation, we should all give thanks for the change they created. Because of them, we can only imagine the horror of going without food, and shall never know what it is to see our children starve. Perhaps if modest increases in weight are the price of winning that fight, it is worth paying.

That said, even if obesity is just a side effect of general improvements in our food supply, if it presents a grave threat to people's health, then it does still need to be addressed. But for obesity to be considered an epidemic, there needs to be evidence of widespread harm being caused, unless someone is willing to admit that they simply don't like looking at fat people.

Changes in BMI are only significant if crossing a particular threshold is equivalent to contracting a harmful or life threatening disease. And while most people accept that there are health risks associated with carrying around large amounts of fat, in the next few chapters we shall discover that even this is not quite as simple as it seems.

5

HOW UNHEALTHY IS
IT TO BE FAT?

This is the first generation of kids expected to die before their parents.
Jamie Oliver

*[T]his crisis marks the first time in our history that a generation of
American children may face a shorter expected lifespan than their
parents.*
From Michelle Obama's Partnership for a Healthier America
website

*Because of diabetes and all the other health problems that accom-
pany obesity, today's children may turn out to be the first generation of Americans whose life expectancy will actually be shorter
than that of their parents.*
Michael Pollen

*These were just back of the envelope plausible scenarios. We never
meant for them to be portrayed as precise.*
David Allison, co-author of the much-covered
'Special Report' on which the above quotes were based

A DYING GENERATION?

Obesity is big news. Supposedly, this is because being fat comes with a number of known health risks, so the endless coverage, the disembodied fat torso pictures, the gratuitously wobbling bellies can all be justified as being for the good of the nation's health. Of course, for the media's purposes, the fact that we find fat people both fascinating and disgusting helps make obesity a compelling story anyway, but without the huge health burden as an angle, we would have to confront some pretty shallow truths about ourselves.

As part of this coverage, the claim detailed in the quotes above, that this generation of children will die before their parents, is often repeated. This is not surprising, because it is a shocking indictment of the seriousness of the epidemic. To this day you will hear academics and campaigners repeating it as a known fact, yet it was not based on any research, and was extrapolated from comments made in a 2005 'Special Report' in the *New England Journal of Medicine*. The report contained no statistics to support the claim, and the authors, one of whom is quoted above, later had to admit it was based on various assumptions. In contrast, the World Health Organization predicts, from extensive analysis of global population data, that life expectancy will continue to rise.

Similarly, statistics from a 1996 paper suggesting that 300,000 Americans die every year from obesity have been thoroughly debunked by scientists at the CDC who have suggested that this figure is nearly three times too high.[1] Yet the 300,000 number continues to be rolled out at conferences and talks around the world, and endlessly repeated in the media. Google '300,000 Americans die from obesity', or 'a generation of children will die from obesity before their parents' right now, and see when the most recent reference is. I guarantee it will be from the last few weeks, quoted by someone who should know better. Both these claims could easily be fact checked and dismissed. Perhaps some commenters even know they are false, yet still feel justified in using them, because their shocking nature might just motivate people into action. There is an epidemic to be fought, and truth, as we know, is an acceptable casualty of war.

I am not saying that obesity is not serious. I am, after all, writing a book about it. One hundred thousand dead Americans is still too many, and there are many known health risks that can be exacerbated by living in an obese body. But we are often more driven by fear than understanding.

IS YOUR FAT KILLING YOU?

It was the American insurance industry that first started to pull together data on body weight and health. There had long been an assumption that excess weight came with a number of risks, but it was not until the 1940s, when Louis Dublin, the chief statistician at the Metropolitan Life Assurance Company, produced tables of 'desirable' weights based on vast sets of North American data, that health professionals had any clear guidelines on what their patients should weigh. In 1960, MetLife published a report updating these figures, indicating that mortality was lowest for people weighing a little less than the US average at the time. MetLife tables were pinned to the wall of every doctor's office, showing that the risk of death rises steadily as people grow fatter, thus embedding an association between thinness and health firmly into the American psyche. And as the ideal weight was below the average, that meant that for the first time ever, the majority of people were being told to lose some pounds.

Pretty much ever since, this correlation has been widely accepted, creating a fear of fat within the medical community, and a burning desire to lose weight among pretty much everybody else. The consensus was formed that any level of thinness was desirable, something that came to dominate US cultural attitudes for many years. It is tempting, although perhaps a little simplistic, to speculate that this led to the cultural idolisation of thin, waiflike women, with the curvaceous sirens of the 1940s and '50s morphing into the skinny, sometimes emaciated, forms that dominate catwalks today.

THE STRANGE CASE OF THE U-SHAPED CURVE

In 1980, having recently been appointed as the clinical director of the National Institute on Aging, the gerontologist Dr Reubin Andres was attending a conference on obesity. He was struck by how much was being taken for granted, particularly how MetLife data from the 1940s was still being used to assess risk. He started to undertake some investigations of his own and became more and more convinced that something was not right. He reviewed forty studies from around the world, and great swathes of US actuarial data, and discovered that the lowest rates of death were found in people between 10 per cent and 20 per cent heavier than the MetLife recommendations. Crucially, he also discovered that as people aged, higher weights seemed to be even more strongly associated with increased longevity. The results, when broken into age groups, produced a series of U-shaped curves, with the U becoming more pronounced as people's age increased. At very low weights, mortality was high, dropping down to a lower mortality for those who were 'overweight', then rising again as people became heavier still. As people aged, there seemed to be an increasingly large risk associated with being underweight. And as they got older, the ideal weight for minimising the chance of death rose. For every decade beyond your forties, a gain of around 3 kilograms seemed ideal.

For Andres, the message from the data was clear. For a long life, people should start off fairly thin, and gain a small amount of weight every year. The MetLife data had made the mistake of focusing on people's ideal weights at twenty-five, and had missed the protective effect of weight increases as people aged. In an interview with the *New York Times* in 1985, he said, 'For some reason, the idea has grabbed us that the best weight throughout the life span is that of a 20-year-old, but there's just overwhelming evidence now that as you go through life, it's in your best interests to lay down some fat . . . Most of us go by aesthetics because we just don't like the look of fat people. But the best aesthetic weight may not be what is desirable if you want to live longer.'

His work proved hugely controversial. Since the 1960s, aesthetic and medical ideals had coincided, but here was data showing something quite different. At around the same time as Andres's study, Ancel Keys produced a review of data from 12,000 men, showing that in the absence of high blood pressure, being overweight was not a risk factor for mortality. This, from the most prominent nutritionist of his day, seemed to back up some of Andres's claims, but still the medical community closed ranks. They protected the use of their precious MetLife tables, and continued with their constant vilification of fat.

To be fair, there were difficulties involved with recommending moderate weight gain as people aged, advice that might have the potential to confuse the public. But I also suspect that Andres's work was so controversial because it challenged one of our great assumptions that the bodies we see on billboards are a genuine marker of good health. Much as a 'healthy tan' is now known to be an oxymoron, perhaps our idea of what looked like a 'healthy weight' was equally flawed.

Andres stuck to his guns, convinced that he was correct, and deeply troubled by the intractability of a healthcare community wedded to the aesthetic of thinness. There were doubtless holes to be picked in his data, with many commenting that it didn't account for weight loss through undiagnosed illness. He also failed to take into account the effects of smoking, something that tends to result in people dying earlier and leaner, probably making the higher weights seem healthier. But if nothing else, his challenging of the consensus should have underlined the difficulties in nailing down associations between weight and health, echoing the comment made by Ancel Keys a few years before that value judgements should never be made based on someone's weight.

Andres's work was largely forgotten, but in 2005, Katherine Flegal published a large study looking at the associations between BMI and mortality in the US. This was considered at the time to be a significant piece of work because, unlike Andres, Flegal was far from being a lone outsider. The report was produced by the National Center for Health Statistics, part of the CDC, where Flegal was senior scientist. As a government scientist, she was not allowed to take a salary from any corporations, freeing her from accusations of industry bias. On top of that, within the world of epidemiology, Flegal is considered one of the leading figures, among the most respected and cited researchers in her field (although she has modestly claimed in the past that she is only cited so much because of the huge amount of research published on obesity, almost all of which cites her 1998 study).

Surprisingly for many, Flegal's study seemed to echo the conclusions of Andres, indicating that being in the overweight category, with a BMI between 25 and 30, was associated with a lower risk of dying than the 'normal' weight category of 20 to 25. It even indicated that being obese, with a BMI between 30 and 35, only had a weak association with mortality; so weak, in fact, that it was not considered statistically significant. Crucially, Flegal's work was detailed and thorough and had accounted for the effects of weight loss due to smoking and illness. And coming from one of the most respected professionals in the field, it was going to be far harder for people to find flaws in it.

The study also found that being underweight was associated with a hugely increased risk of death, as was having Class II and III obesity, representing a BMI of over 35. Like Andres, the numbers produced a U-shaped curve, with the bottom of the U sitting a lot further to the right than many people thought it should. And also like Andres, Flegal showed that these effects seemed to become more pronounced as people got older, indicating that getting a bit fatter as you age might be good for you.

The study also showed that the association between obesity and early death had been decreasing quite significantly over time, with the suggestion that this might be due to improved medical care for certain serious conditions. Given that life expectancies in the US had been improving while people's weight had been going up, this could perhaps have been expected, but strangely it was this finding that really upset people. It seemed to indicate that the health impacts of obesity were becoming less serious as time went on, a message that many thought was too dangerous for the public to hear. Perversely, many in the medical community refused to celebrate the fact that fat people were living longer.

In fact, the findings caused much consternation. Previous studies had shown far higher associations between obesity and mortality, and a high quality investigation showing figures dropping caused shock waves around the world.[2][3] The adverse reaction seemed to surprise Flegal, a scientist who had always claimed she just wanted to report the data as she saw it. She was perhaps not as aware as Andres of the deeply entrenched views people have when it comes to fat, or if she was, she didn't want to be drawn into the debate. A few years later, she was quoted as saying, 'You can't really say that they had data that disagreed. You'd have to ask them exactly why they didn't like it. It was just kind of a strong reaction . . . People just seemed to like the higher estimates. People seem to like high estimates of everything. They see it as something you can use to motivate people. Our estimates dropped the numbers really by a lot. That took people aback.'[4]

In the years that followed, a number of other researchers seemed to confirm Flegal's findings, and so in 2012 she decided to perform a meta-analysis to bring everything together. Her team combined the results of ninety-seven studies,[5] publishing early in 2013. With such an enormous data set, just under three million participants, the feeling was that it should give a definitive answer to the associations between obesity and mortality. Once again, the study showed that people in the overweight category were less likely to die than those of supposedly 'normal' weight, and that there was no association between Class I obesity and death. Underweight categories, and those with BMIs over 35, did show a much greater risk of dying, again producing a pronounced U-shaped mortality curve, with the bottom of the U sitting right in the current overweight category. Here was a huge study, by one of the field's most respected academics, and it seemed to show that being 'overweight' was not likely to kill you. Even living in the Class I category for obesity did not seem to carry an increased risk of death.

The reaction was nothing short of vitriolic. One of the most vocal critics was the Harvard professor of nutrition and epidemiology Walter Willett, someone who had also been a vociferous critic of Andres's work many years before. He took the bold step of organising a public symposium to debate Flegal's meta-analysis, claiming that the paper was dangerous and would confuse the public. In a radio interview he famously said that 'this study is a pile of rubbish, and no one should waste their time reading it'. This view was an early indication that the symposium was going to be more of a demolition job than a debate. Experts lined up to point out methodological flaws in the data, and Willett, a popular media commenter on matters of nutrition, made sure his views were widely heard.

But some of his comments seemed to have gone too far for the world of epidemiology, especially when aimed at the well-respected Flegal. In the years before, evidence had been growing to support the idea of a so-called obesity paradox, the suggestion that a high BMI might actually be associated with greater survival rates from certain chronic conditions, particularly cancer and cardiovascular disease. Many pointed out that there were plausible mechanisms that could support the protective effects of weight gain into old age, suggesting that a store of energy could help see people through difficult times.

In an unusual step, *Nature* published an editorial that contained a damning criticism of Willett's attack on Flegal, with some experts suggesting that the Harvard group was interested in suppressing any data that did not fit with the established 'fat is bad' consensus. Willett's claims that the information would confuse the public was countered by Kamyar Kalantar-Zadeh, a nephrologist at the University of

California, who pointed out that 'we are obliged to say what the real truth is'. Many accused Willett of arrogantly not trusting the public with information. Others seemed to be suggesting that the weight of the new information was so great that people needed to be told.

Flegal herself often seems bemused by the criticism and has often claimed that her data is 'not intended to have a message'. But when it comes to obesity, everything is so divisive and laden with judgements, that there is always a message. When John Snow removed the handle of the Broad Street pump, the overwhelming evidence that cholera was caused by something in the water supply, rather than the stench of London's streets, changed scientific opinion almost overnight. But when it comes to fat, things often take a little more time. A BMI over 25 is still classed as 'overweight', which is supposed to be a bad thing even though it probably makes you less likely to die.

And then, just when it seems that the Flegal study had provided the definitive statement on obesity and mortality, a 2016 study from the Global BMI Mortality Collaboration produced a report with conflicting results.[6] It compiled data from over ten million participants that suggested being overweight and obese *is* associated with increased risk of mortality. One might think that a 'Global BMI Mortality Collaboration' has a vested interest in BMI and mortality being linked, and the methods it uses have been criticised,[7] but these conflicting results really underlie how difficult it is to study the links between obesity and death, especially when you are observing people in the real world. There are so many confounding factors, and so much noise in the data, that it is almost impossible to produce a definitive answer.

Even if there is a link between weight and mortality, these are only associative studies, telling us that the two are linked. Does it mean that fat is killing us, or just that people who are fat are more likely to die from other things? Perhaps all that it shows, like ice cream sales and swimming pool drownings, is that some of the things that kill you also make you fat. Or even that the way others treat fat people can be dangerous.

This is the problem about associations. The moment we hear that obesity and dying are linked, it is impossible for us not to create a simple story that pulls them together. We instantly assume that fat is killing us. That our heart is giving out from the effort of pumping blood around a larger body. Or that rogue blobs are blocking up our arteries. That cakes are destroying our liver, or causing cancer. This sort of construction is a dangerous game, leading us to see patterns where there are none, and often missing the real causes.

IT'S NOT ALL ABOUT DYING, YOU KNOW

Of course, it is not all about mortality. The obesity paradox, something that is widely, although not universally, accepted by researchers, refers to fat people being less likely to die from certain diseases. But that does not mean that fat people are less likely to develop those conditions – in fact, in many cases the opposite seems to be true. Many of us would trade a higher risk of death for a better quality of life. This, combined with continued uncertainty about the actual risks of mortality, is often used to justify the continued public health insistence on getting people into the normal weight category.

This carries with it the rarely discussed implication that people are being told to trade off the chance of a longer life, for one that is shorter but free from illness. The dream of all public health professionals is that people live healthily to a reasonable age, and then drop dead quickly and cheaply, having had little engagement with healthcare providers. The lives of fat people in the modern world are at odds with this. They seem to be living longer, and are more likely to survive complex conditions, which makes them extremely inconvenient. This could be one of the reasons why these increases in survival and mortality are rarely celebrated or discussed, hiding the information from the public for the good of an underfunded healthcare system. And although there are very real pressures on almost all healthcare budgets around the world, one might think that it is sensible to allow individuals to make the choice between longevity and freedom from illness for themselves. Some might still choose the extra years, if they knew they had a choice.

HEALTHISM

The idea that a lower chance of getting sick should be inherently preferable to a lower risk of dying early can perhaps be traced back to the concept of healthism, originally described by the political economist Robert Crawford in 1980. He observed a shift in mentality when it came to health, as the rise of neoliberal ideologies cleverly moved the blame for sickness and disease onto the individual. A good citizen was a healthy one, someone who did everything possible to be free from disease, desperate not to be a burden on society. Many trace the rise of these ideas further back, to the work of the pioneering British epidemiologist Geoffrey Rose, who championed population-wide prevention strategies to address specific conditions or risks – changing the behaviours of many in order to save a few. The wearing of seatbelts in cars, for instance, is something that millions of individuals are required to do in order to save the small number involved in accidents. It was a simple shift in behaviour with clear benefits. Yet the simple cause and effect when it came to seatbelts, and the lack of unintended consequences, was very different to the complexity underlying most chronic health conditions.

Perhaps unwittingly, Rose's work paved the way for the belief that good health was the responsibility of the individual. Advertising, political rhetoric and the rise of neoliberal values meant that being healthy became your own responsibility, and this was to become deeply internalised within our culture. Being a burden on the health system was positioned as something that could be easily avoided with good moral choices. Never mind the overwhelming evidence that two people could live the same lives, and yet experience remarkably different health outcomes. Never mind that most medical conditions develop through a combination of genetics and luck. Sick citizens were failing us all.

As obesity rose, and increasing evidence of a link to common diseases emerged, there was suddenly a convenient outward symbol of irresponsibility. Fat people carried evidence of their sins around with them, allowing us to judge them from a distance and assess how much of a burden they were. Healthism made fat people public enemy number one.

But as well as making sickness our fault, healthism also meant that being sick was a choice. In order to motivate people to act, sickness had to be shamed. The idea that a significant or valuable life could be lived within a framework of illness or disability was lost in the relentless pursuit of healthism. Sick people were no longer a vulnerable part of society we wanted to help – they were a deviant minority that we needed to change. Strangely, this is the same set of values that gave rise to the wellness industry, and the many 'food as medicine' health bloggers I wrote about in my first book. Both frame illness as the failure of the individual. A good citizen pays their taxes, eats only carrots, and has the decency to drop dead the moment they are ill.

THE RISKS

There is little doubt that being overweight and obese is associated with a large number of different medical conditions, many of which can hugely impact life quality. There is not much value in exhaustively listing the many conditions thought to have an increased risk due to obesity, so for now I shall briefly discuss four of the most significant.

1. Cardiovascular disease

Cardiovascular disease (CVD) refers to any disease that affects the heart or blood vessels, including heart attacks and strokes. It is usually associated with blood clots and a build up of fatty deposits inside our arteries. The exact causes are not always clear, although the majority of CVD events are probably due to a number of different genetic and lifestyle factors interacting. Risk is increased when people have high blood pressure, high blood cholesterol or type 2 diabetes. Alcohol, smoking, poor diet quality, ageing, genetic factors and family history are also known to be large contributors.

Obesity is considered a risk factor largely because of its association with high blood pressure, high blood cholesterol and type 2 diabetes. Overweight and obese people are frequently told that they should lose weight to reduce their risk of dying of cardiovascular events, but in reality, general prevention advice includes eating a varied and balanced diet, cutting down on saturated fat and getting some exercise.

Coronary heart disease is the deadliest form of CVD. It is the UK's biggest killer, and the leading cause of death worldwide. It causes 70,000 deaths in Britain every year, the majority of which are as a result of heart attacks. Strokes account for another 40,000 deaths, and 240,000 hospital visits.

Despite these terrifying numbers, the treatment and prevention of all forms of cardiovascular disease has seen a great deal of progress in recent years. Incidence of all forms has decreased markedly since the 1980s, with mortality rates also falling dramatically. In the 1960s, seven out of ten heart attacks were fatal. Today, that figure is down to three out of ten.

But CVD still takes a huge toll on our population, cutting lives short and using an extraordinary proportion of our overall healthcare resources. As a result, any behaviours that can slightly reduce population-level risks are likely to end up saving many lives, which perhaps explains much of the focus on reduction of obesity levels. Many people think that population-level increases in BMI have the potential to reverse much of the recent progress made in prevention and treatment of CVD.

We shall discuss cardiovascular disease a number of times as we go on, and look more closely at some of the many risk factors. But it is well worth noting that there are many potential confounding factors when it comes to obesity and CVD, and a number of things that are known to contribute to both conditions independently. There are also a good number of CVD risk factors that are much easier to address than people's weight, perhaps indicating that relentlessly telling people that they need to become thin might not always be the best strategy.

2. Cancer

Several types of cancer are more common in people who are overweight, and Cancer Research UK has suggested that one in every twenty cases of cancer are linked to bodily fat. Jessica Kirby, CRUK's senior health information manager, told me more about this risk:

> There is strong evidence that thirteen different types of cancer are caused by obesity, including two of the most common, breast and bowel cancer, and three of the hardest to treat, pancreatic, oesophageal and gallbladder cancer. There are a number of ways obesity is thought to cause cancer. Fat cells pump out hormones and other growth factors, and extra body fat can also cause inflammation. All of these things can tell cells to divide more often, increasing the chance of cancer cells developing. But research is continuing to understand more about all the ways in which obesity can cause cancer.

This has led some to suggest that obesity is the second largest preventable cause of cancer after smoking. And even though the evidence is largely associative, that does not mean it should be discounted. After all, associative links between lung cancer and smoking were enough to establish that as a cause.

But there is also evidence that the relationship between obesity and cancer is complex and nuanced. For instance, it seems that being overweight is protective when it comes to lung cancer,[8] especially in smokers. As lung cancer is still the most commonly fatal cancer by some distance, accounting for one fifth of all UK cancer deaths, advice to lose weight may not be protective in every case. Of course, stopping smoking would be a far more effective and advisable strategy than putting on weight for anyone worried about lung cancer, but it does demonstrate what a tricky and thorny issue obesity can be.

Cancer Research UK is keen to raise awareness of the cancer risk from obesity, as its evidence shows that many people are not aware of the links. But this does create an interesting conundrum. Is a raised awareness of specific disease risk likely to be a motivating factor for people struggling to lose weight? Are obese people sitting at home unaware that they are damaging their health, just waiting until cancer is brought into play before they change their errant ways? Or are they already terrified and ashamed of their poor health, yet still unable to simply become thin?

Some might say that a little more awareness could provide the ultimate motivational push, and help people finally lose weight. But for me, the big issue is that obesity is not a behaviour like smoking. The greatest cancer risk of all is getting old, yet there is no drive to raise awareness of that, largely because this is something we cannot change. I do not wish to denigrate the work of CRUK – it does amazing and positive things – but I am not convinced that making fat people even more afraid, and linking their size to one of the most stigmatised of all health conditions, is the most effective way to help them.

3. Type 2 diabetes

When it comes to obesity and health, the condition that generates the most apocalyptic predictions is type 2 diabetes. We are forever being told that it is set to send us spiralling into oblivion, likely to bring the worldwide provision of healthcare to its knees.

Type 2 diabetes accounts for around 90 per cent of all diabetes cases, and is caused when our body develops insulin resistance and a fall in insulin production. We shall cover insulin and diabetes more thoroughly in Chapter 13, but briefly, insulin is an important hormone that helps control our blood glucose levels. If we develop insulin resistance, our body stops responding to insulin when it is released, and so we cannot control levels of glucose in our blood. Uncontrolled blood glucose levels can cause extensive nerve and blood vessel damage, leading to a number of severe health complications, including cardiovascular disease, blindness, kidney failure, limb amputations and early death. Once type 2 diabetes has developed, it needs long-term treatment with a combination of dietary and drug therapies. Often this is expensive and challenging, with management usually required for the rest of people's lives.

Although the causes are unclear, recent evidence seems to suggest that the body's inability to respond to insulin is caused by high levels of fatty acids circulating in the blood.[9] One hypothesis is that our fat cells only have a limited capacity for storage, and once that level is reached, fat has nowhere else to go, so ending up in our blood and interfering with the processes that allow insulin to control glucose levels. This might account for why some people, particularly South Asian and Chinese populations, develop type 2 diabetes at relatively low weights, as perhaps their fat cells have a lower storage capacity. Of course, in reality, this is probably a vast oversimplification of a complex condition, which is likely to have many potential causes. Insulin resistance alone is not the whole picture, as in many cases, a drop in insulin production is a far more significant cause.

Regardless of why it occurs, type 2 diabetes is a brutal and devastating condition, and something strongly associated with obesity. Most people consider that the twin rises of these conditions in the past thirty to forty years is no coincidence. Between 1985 and 2015, worldwide incidence of type 2 diabetes has risen from 30 million to nearly 400 million cases.[10] [11] Although there are many potential reasons for this extraordinary rise, including an ageing population, longer life expectancies, increased recognition and diagnosis, poor-quality diets and falling rates of exercise, one of the biggest factors increasing risk is thought to be obesity.

But as we shall discuss throughout the book, the causes of this condition are not well understood, and the relationship to obesity is thought by many to be multidirectional. Some consider obesity to be an early symptom of diabetes, so perhaps not a cause at all. Others point the finger at drugs used to treat the condition, many of which can cause weight gain and so provide a confounding factor.[12]

But one thing is for sure. Type 2 diabetes is one of the greatest health threats facing the modern world. In 2017, it was estimated to have caused four million deaths. Crucially, it is an increasing threat, with rates rising in most countries. Although we are getting better at treating it, we seem to be unable to do much to prevent its rise. Treatments are complicated and expensive, placing a very heavy burden on healthcare systems everywhere. And the personal cost is often devastating, greatly affecting a sufferer's life quality and risk of death.

But any suggestion that type 2 diabetes is a result of people letting themselves get too fat through bad choices is overly simplistic at best. Increasingly, this condition has become a stick with which to beat fat people, a way of demonstrating how selfish and irresponsible they are. Type 2 diabetes is a concern for health systems, and often a crisis for individuals, but ignorance, stigmatisation and blame won't make things any better. Right now we are freely blaming people, their lifestyles and the systems that feed them, and insisting on wholesale changes. But there are always dangers when we do this without fully understanding exactly what is going on.

4. Metabolic syndrome

Metabolic syndrome refers to a cluster of related conditions that increase health and mortality risks. Someone is diagnosed as having metabolic syndrome if any three of the following five factors are present:

- central obesity (defined by measuring waist circumference, not BMI)
- high levels of fatty acids in the blood
- low levels of HDL (the so-called good cholesterol)
- high blood pressure
- raised fasting blood glucose (which would indicate insulin resistance).

As many of these factors are interlinked, it is perhaps unsurprising that metabolic syndrome affects a large number of people, up to one in four

UK adults. It provides a way of combining factors together to help identify those most at risk, and so recommend lifestyle interventions or treatments. It is telling that central obesity is used rather than BMI, because this is thought to more accurately correlate to risk. Waist circumference is extremely simple to measure, and much easier to calculate than BMI, which does perhaps raise questions as to why BMI is favoured in so many other contexts.

Clearly, being obese or overweight increases your risk of being diagnosed with metabolic syndrome, which in turn does seem to be a good predictor of future health outcomes. People diagnosed with it are five times more likely to develop type 2 diabetes, twice as likely to suffer from cardiovascular disease, and twice as likely to die from it than other sufferers of CVD.[13] The main interventions recommended for those diagnosed are simple lifestyle changes such as a healthy diet, regular exercise and stress reduction. Sometimes drug treatments are used if these aren't working.

5. *Other things*

Although those are the four main health problems associated with obesity, there are also a number of others worth mentioning briefly. People may suffer mobility problems, especially in the most severe cases, and clearly excessive weight gain can have a severe impact on people's life quality. Carrying around a great deal of weight can lead to wear and tear on joints and bones, and sometimes the world is a hard place for larger people to navigate. Some of this is a direct effect of their weight, but some is a consequence of a poorly designed environment that does not suit everyone. It would be considered shameful and ridiculous if we did not adequately cater for the population increases in height we have seen over the past 100 years, so it is curious that we are rarely keen to do the same when it comes to weight. (Imagine rolling your eyes when someone stoops under a low ceiling.)

There are also connections between obesity and a number of mental health conditions, including depression, low self-esteem, emotional issues and behavioural problems in children. Suffice to say there is no known mechanism by which bodily fat might physically cause someone to experience mental health problems. So we can assume that, at least in this case, these serious issues are due to the way that fat people are treated, not some physical effect of their BMI.

No one would suggest tackling the social harms of racism by painting black people white, so it is hard not to think that a great deal of this might be better addressed by changing societal attitudes, rather than insisting people shrink.

SOME REASONS TO BE SCEPTICAL

The main issue with any of the health conditions linked to obesity is that the associations being made are largely observational. With the exception of some cancers, there is very little definitive evidence of mechanisms by which excess weight actually causes serious medical conditions, and so the majority of the associations remain just that. But when it comes to obesity, causal relationships are frequently implied in a way that is rarely done in other areas, perhaps driven by our cultural biases and beliefs. It is not uncommon to hear declarations that obesity causes all of the conditions listed above, and many more, with people failing to appreciate that an association with increased risk is not always evidence of cause and effect. Obesity and illness happen out in the real world, where things are messy and confusing. It is often very hard to pick apart whether or not one thing causes another. Some things that make you fat also make you sick. And some types of illness can also make you gain weight.

To be clear, I am certainly not suggesting that the links between obesity and disease should be discounted, especially when it comes to the most severe cases. Extremely obese people often suffer tremendously, yet it is important to understand properly what the causes of this suffering are, and what actions will genuinely help them. When body weight is the laser focus, there is a chance that we might miss ways of helping people improve their health, especially given the long-term ineffectiveness of most weight loss approaches.

If fat is a cause of serious health conditions, then it would perhaps be expected that fat removal might result in the risk of these diseases dropping. In many cases that does appear to happen, but of course it becomes very hard to separate fat loss from the behaviours that people engage in to cause it. Improvements to diet, increased levels of exercise and improved self-esteem might well cause many of the health changes frequently attributed to a loss of weight. When liposuction patients are studied after having a significant amount of fat sucked out of them, there does not seem to be any health benefit. Yet when people lose the same amount of weight with improvements to their diet or increased activity, their health does seem to improve. Additionally, most of the benefits of weight loss seem to be independent of how many pounds someone has shed,[14] which does suggest that the behaviour changes might be more significant than the amount of fat lost.

NOT ALL FAT IS EQUAL

To further complicate things, the type of fat present might actually be more significant than the amount, something that simple measures of BMI do not distinguish. Diet targets are based on the numbers of pounds lost, and anyone moving into a lower BMI classification will feel disproportionately happy about the change they have made. But all bodily fat is not created equal.

One of the key factors is location. Visceral fat, located around the waist and the abdominal cavity, is thought to be far more harmful than fat around the hips and thighs, leading many to argue that we should focus on waist circumference rather than BMI.[15] It is certainly true that this would provide a far better indicator of health risk, but there are practical issues. If someone has a large waist, it can be hard to measure without exposing the navel, and can be quite error prone if the waist position is hard to define. Such difficult measurements involving a lot of physical contact might be humiliating to large patients, and despite the inherent bias in healthcare, the vast majority of medical professionals do try to show compassion and understanding. BMI is also easier to measure in the elderly or infirm, as it can still be assessed with patients in a seated position. (Arm span is roughly similar to height in most people, so can be easily measured even if patients cannot stand.)

But it is also tempting to surmise that BMI remains the main diagnosis tool because it has been that way for so long, and can be easily compared to old databases to see how things have changed. It is also a possibility that BMI is popular because it shows better correlation to our aesthetically driven perceptions of what healthy looks like.

One of the reasons why waist circumference shows better correlation to health than BMI is because it is a better indicator of how much visceral fat someone has – a term that refers to fat deposited around the liver and kidneys. Broadly speaking, increases in visceral fat are harmful to our health, whereas subcutaneous fat, the fat located under our skin, is relatively harmless. Exercise and quality of diet can influence our amount of visceral fat, and so provide lots of health benefits independent of weight loss.[16] So although a patient's BMI might stubbornly remain unmoved, the actual improvements to their health might be real and measurable.

This is one of the great problems with weight loss dieting. Subcutaneous fat is the fat we see, the type that causes us so much stress, discomfort and prejudice. Yet this fat is relatively benign, and may be having little impact on our health. When the scales fail to move even when the dieter is doing everything they were asked, it is only natural to lose motivation and abandon the attempt. It can lead to people giving up on a course of action that might be genuinely helping.

The reasons why these different types of fat have different effects is perhaps not as obvious as you might think. Dr Giles Yeo is a principal research associate at the University of Cambridge and President of the British Society for Neuroendocrinology. He told me, 'Where you store fat is important, whether it is visceral or subcutaneous. But the differences are not because of the geography. If you take visceral fat from a mouse and transplant it into its butt, you get the same sort of metabolic dysfunction. Different fats produce different hormones and have different effects.'

Visceral fat is thought to release inflammatory compounds when present in large amounts, leading to a number of health problems. But within visceral fat, there is thought to be a great deal of variation. Different types produce varying levels of inflammation, and having large amounts of the most damaging types is strongly associated with insulin resistance.

Even though numbers of fat cells in adults are generally fixed, there is some evidence that external factors might affect them. When mice are fed a high fat diet, new adipocytes grow in their visceral fat, but not in the fat under their skin, indicating that the uncontrolled creation of new fat cells might lead to metabolic harm.

Fat is too complex to pigeonhole as either bad or good. But for some reason, loss of total fat is still seen as the ultimate goal of all dieting, and reductions in BMI as the only way we can improve population health. In truth, it is not fat that we should be fearing, but the damage and disease often attributed to it. Improving your diet and exercising may have mixed results when it comes to losing weight, but if we are talking about improving your health, they can be very effective, particularly when the alternatives are often so extreme.

BEING BULLIED IS BAD FOR YOU

When we discuss the links between obesity and disease, one thing that is very rarely talked about is the stigma and social pressures that come with a socially undesirable body shape. A 2015 study by the psychology researcher Dr Angelina Sutin found that weight stigma was associated with a hugely increased mortality risk.[17]

The psychology researcher Dr Eric Robinson studies the effects of weight stigma on obese people at Liverpool University and told me this:

> We know that obesity is associated with ill health, but the question is, how much of that is physiology, and how much is psychology? The psycho-social experience of being obese definitely seems to increase the chance of developing illness, but as yet we don't know how much. Certainly, most people with obesity seem to have experienced some sort of mistreatment, which is likely to have an effect. The results on increased mortality are absolutely shocking, but for some reason are rarely talked about, even in academia . . . It is actually hard to separate the objective and subjective state of obesity. At the moment we are looking at experiments where we take people of normal weight and make them wear a body prosthetic so they appear obese, then send them out into the world. It is a way of making them feel the stigma of obesity. So far the data looks interesting. They have impaired self-control and overeat energy dense snacks. It appears that the psychological consequences can manifest as negative behaviours.

Dr Robinson's team has also looked at how someone perceiving themselves as overweight might influence future health outcomes. You might assume that knowledge is power when it comes to obesity, and this assumption underlies the UK's childhood measurement programme. But it appears that telling people they are overweight might not be helpful. Dr Ashleigh Haynes has studied how people who self-identify as overweight or obese fare against those who are equally fat, but falsely see themselves as normal weight. The results should give us pause.[18] [19] Dr Haynes told me:

> There is a certainly a view that people need to be made aware of their weight status, which underlies many public health approaches. We looked at data from the last twenty-five years to try and understand the relationship between self-perception and weight management behaviours and the data is pretty consistent. Unsurprisingly, when people think they are over-weight, they will try and lose weight, but there is no consistent link to them eating healthily, or an increase in physical activity. People who perceived themselves as overweight were more inclined to disordered eating, and it was also a predictor of future weight gain. It is difficult to know how this translates into advice for people and health professionals. Is it better not to tell someone they are overweight given the widespread stigma asso-ciated with larger bodies?

Studies have shown that experiencing weight stigma is associated with an increased risk of depression,[20] stress, binge-eating disorder, anxiety and a reluctance to engage with healthcare. There is also evidence that weight stigma may cause increases in weight gain and be associated with a number of the health burdens of increased weight, including cardiovascular disease.[21] [22] Fat shaming is often sanctioned under the perverse pretence that it is motivated by compassion, but these actions seem to be making people fatter and sicker. Stigma goes way beyond a few objectionable media commenters. It comes from government-issued leaflets and reports, television documentaries and weight-loss reality shows. But most of all, it exists in every person who believes fatness has a simple cure.

Thankfully, we are beginning to wake up to the problem. For her thoughts on this I spoke to Susan Jebb, who advises the UK government on obesity, and makes recommendations for healthcare professionals.

> I am the first person to say that there is shocking discrimination against people who are obese in society at large and doctors are not necessarily immune from such views. I try increasingly hard

to prevent the messages we give out from contributing to stigma, but can't promise I always get it right. Some specific things I try to do are always use people first language and non-stigmatising images – we have a bank of images of people who are overweight that we use and encourage others (including the media) to use which show people who are overweight doing ordinary things, to try to escape from the images of lying on sofas and eating burgers. In presentations to students I discuss the issue of weight stigma and how unconscious bias can affect our relationship with patients.

In order to help people, the focus must fall on improving people's lives, and not on fighting off an imagined fat enemy, freeing the thin people trapped within. We must show concern for people's health, not for their appearance.

Occasionally, that will require us to accept that some people are always going to be fat, and help them live their lives in the best way they can. Every life comes with individual risks. Our genes, our lifestyle, our decisions, and the inevitable act of ageing, all contribute hugely to the chance that we might become ill at some fuss point. And yet, perhaps with the exception of smoking, there is no single risk factor that seems to prompt as much judgement as being fat. This is perhaps all the more remarkable because it is, to a very large extent, just how people are.

The health burden is definitely real, particularly where BMIs rise above 40, but tragically, this is often used as a stick to beat fat people with, rather than a reason to feel empathy and compassion. Are we really worried about their health, or do they just disgust us? Because if all we want is for them to be well, perhaps we shouldn't treat them so badly.

THE MAGIC PILL

Just imagine for a moment that a drug was developed tomorrow that removed all the negative health implications of being fat. A pill that would completely sever the links between fatness and chronic disease, and simultaneously address any related mobility issues. If this was somehow achieved, would that be the end of all the fuss? Would all calls for people to lose weight abate? Would I no longer be afraid of getting fat? Would the diet industry cease to exist? Would the world create an environment that better suited people of a larger size? Would airlines and restaurants install larger seats? Would the media end its vitriolic attacks? Would the next prime minister refrain from commenting?

As Reubin Andres noted nearly forty years ago, we are wedded to the idea that what we find pleasing to our eye must be representative of good health and vitality, and anything that challenges that is deeply troubling to our minds. This is despite the knowledge that what visually pleases us is a product of our culture.

But this magic pill does not exist, and the fact remains that there is still a significant health burden associated with obesity. People leading longer lives, with a greater chance of becoming sick, is a problem – not just for the individuals concerned. Our healthcare systems need to have the resources to treat us, and many of the health conditions larger people are at risk of are extremely expensive to treat.

As a result, we are forever being told that obesity has us teetering on the brink of a financial disaster, about to drag us into the abyss. But how true is that? Just how much does obesity cost?

6

HOW MUCH DOES
OBESITY COST?

*Seeing care for certain groups as an excessive cost reflects an argu-
ably perverse way of thinking about healthcare in terms of human
need . . . In other words, care for the sick is an economic burden
only in healthcare systems where profit is the bottom line and
public services are underfunded and politically unsupported – that
is, systems in which only market logic is considered legitimate.*
Julie Guthman, *Weighing In: Obesity,
Food Justice, and the Limits of Capitalism*

Obesity is expensive. Exactly how expensive is hard to measure, but rises
in body weight certainly come with a cost. There are the costs of treating
obesity-related disease, which can run into billions. It is estimated that
in developed nations, between 2 and 7 per cent of all healthcare budgets
are attributable to conditions caused by obesity. The UK government's
2007 Foresight report suggested that in the UK the total annual cost of
obesity in 2002 was around £7 billion, with £1 billion attributed to
healthcare. Other costs included lost earnings, lowered productivity and
welfare payments. A later study in 2011 indicated that the healthcare
figure was a little over £5 billion.[1] Then, in 2014, a report from the
McKinsey Global Institute reported an extraordinary healthcare-only
figure of £16 billion. These numbers did not represent rapid rises in
costs, simply different ways of calculating the financial toll.

Perhaps the main issue with all attempted calculations is that obesity
is not actually a disease. So when figures for 'obesity-related conditions'

are reported, these do not communicate the cost of obesity at all. For instance, diabetes is a very expensive condition to treat, and it is estimated that it costs the NHS around £10 billion annually. But a significant proportion of these cases are not attributable to obesity, and would have happened whether or not people were fat. In order to work out the actual cost of obesity-related diabetes, we need to understand what proportion of cases are attributable to that alone, and then work out how much those cases cost. The same is true for heart disease, cancer, and all the other expensive health conditions associated with being fat. There is much margin for error in this sort of calculation, and, as a result, different methods of calculation can result in estimates an order of magnitude apart, even when taken from the same data.

Much as with mortality rates, sometimes figures are adjusted in order to increase their shock value. As Katherine Flegal wisely noted, people always seem to like higher estimates of everything. The widely reported £16 billion figure turned out to be the 2011 £5 billion figure adjusted for inflation, added to the £10 billion figure for the treatment of all diabetes. This meant that not only did it count the cost of obesity-related diabetes twice, but also added on the cost of all diabetes not related to weight. The £10 billion figure even includes type 1 diabetes, a condition that is not related to obesity at all.

As a result of all these confusions, the true cost of weight gain is hazy and hard to measure. Almost all economic calculations about obesity assume a causal relationship when there is just association, because this pushes the estimates far higher.

PARTIAL EQUILIBRIUM

Perhaps the biggest problem with any calculation of healthcare costs is that all models use a 'partial equilibrium' approach to their calculations, ignoring the impact that changes will have on other parts of the health service, or indeed other sectors of the economy.[2] When it comes to healthcare spending, the idea that costs are unnecessary and avoidable is, quite frankly, absurd. Everyone is destined to die somehow, usually after a period of expensive sickness, and removing one cause will only lead to another. Unless you can definitively say that the alternative cause of demise will be cheaper, the savings made by avoiding one cause of death are irrelevant. Often, as people age, the conditions they face become increasingly expensive, requiring long-term care and treatments, but just because it's expensive doesn't mean we are going to stop trying to help people live longer.

By far the greatest burden on all healthcare systems is an ageing population. A 1998 paper investigated whether or not eliminating all fatal diseases would have a cost benefit to the Dutch healthcare systems.[3] The report concluded that it would in fact make things more expensive, as people would be more likely to develop degenerative conditions into old age. They wisely concluded that the aim of preventative healthcare should be 'to spare people from misery and death, not to save money on the healthcare system'.

This is the problem with a partial equilibrium approach: it doesn't look at the alternatives. Dying early may be a tragedy for the individual, but it can deliver considerable savings to the economy if that's all you care about. A 2017 study for the Institute of Economic Affairs estimated that the savings from obesity-related early deaths were just over £3 billion, which if taken into account would more than halve the current best estimates for obesity's cost.

At the end of his 2017 paper on interpreting cost of illness studies, the economist Brendan Kennelly noted:

> By focusing solely on costs, one can be distracted from the benefits that are produced from the resources devoted to healthcare. An analogy with the amount spent on entertainment and leisure might be helpful. It is possible (but silly) to think that the total amount spent on entertainment could be saved if only people learned how to be entertained by activities that didn't cost money. Instead, most people think that it is a good thing that individuals and firms have devised all kinds of interesting ways to meet peoples' [sic] demand for entertainment and leisure. I think that most, but not all, healthcare delivered in developed countries represents an appropriate use of resources. We should acknowledge and celebrate that instead of suggesting that those resources could be used for some other benefit, if only certain diseases and problems were prevented.[4]

The cost of obesity-related health problems seems to be assessed in a very different way to that of other conditions, something that seems linked to our belief that obesity is a condition related to a lack of willpower or responsibility. It is also assessed differently to healthcare costs related to drinking, dangerous sports or stress, perhaps revealing something of a class prejudice regarding what the state should pay for.

In reality, the true cost of obesity is perhaps just as impossible to calculate as the health risks, and just as likely to be interpreted to suit a particular agenda. But even if an agreement could be reached, how much value is there in knowing? If there was no obesity, would we all have more money to spend? Would we all live forever? Would our longer lives be cheaper, or more expensive? Would the illnesses that we end up dying of be better, or worse? And given the huge costs, what would be the effect on the economy if obesity were to end? What would be the impact of losing the multibillion-pound diet and weight loss industries? Or the companies that profit from the medical treatment of obesity?

Any attempt to moralise about the cost of fat requires an imagining of a world without it. Every society that has lacked a significant level of obesity has been one where hardship and starvation were rife. Presumably we do not want to return to those times. Surely we will not accept food hardship, and have no desire to see vulnerable people go without.

So just what is it that we want? A world where everyone is the same weight? Given the diversity within humanity, that simply cannot exist. Perhaps we want a world where everyone falls into the normal weight category, despite the uncertainty about what that would do to mortality rates? Despite everybody calling for something to be done, no one seems to know exactly what the end game looks like.

TROUBLE THROUGH THE LOOKING GLASS

Most campaigns and public health interventions are based on Geoffrey Rose's idea that wide-scale interventions are the most effective way to address risk. Interventions are made with the vague intention of curbing rises in BMI, hopefully pushing the bell curve of weight distribution a little way to the left. These have mostly been unsuccessful to date, especially for the most vulnerable groups in society, but if they did end up working, just consider the potential for unintended consequences.

Any successful attempt to shift the curve to the left would be likely to increase the number of people falling into the underweight category.

Katherine Flegal's work revealed the extraordinary links to increased mortality within this group, especially as people age. Her 2013 study showed that being underweight was associated with more deaths than having a BMI of over 40, despite there being many more people in the higher categories. Any dietitian will tell you that their greatest day-to-day challenges involve getting adequate nutrition into sick people. Undernutrition is a hidden problem in our society, largely confined to hospital wards and care homes. It exists in an invisible sick, disabled and elderly population, unable to feed themselves, or with bodies incapable of taking on enough nutrition when they do. The health costs associated with it are shocking, but it is not a crisis that receives much attention from the media or the government.

Undernutrition also has a rarely reported financial burden, and one that is truly appalling in scale. In 2011, the estimated healthcare cost of treating malnutrition in the UK was £15.2 billion,[5] dwarfing even the highest realistic estimates for obesity (although clearly, these too are estimates). Beyond this, try to contemplate the human cost of seeing loved ones waste away from lack of food, an undeserved way of playing out the last years of life. Imagine the cost to employers, with increasing numbers of people pulled into a life of caring for relatives and partners. Think about the cruelty of dying of starvation, and the desperation of healthcare professionals seeing people unable to get enough food into their damaged bodies.

Why is malnutrition not declared a disease, an epidemic, a savagely expensive blight, bringing our health system to its knees? Perhaps because it is complex, the result of many different conditions, and rarely the fault of the individual. Should we really discount the risk that in fighting hard to bring the average weight of the population down, undernutrition might grow? If we want to lower BMIs, we must be clear how we intend to protect the lower weight section of the population from harm. If we take away choice, force people into exercise, shift calorie densities down, tax certain foods, ration others, and continue our drive to make people fear weight gain, what steps will be taken to protect people's health in the new environment this creates?

WHAT WE SHOULD REALLY BE DOING

To tackle obesity, the focus should always be on those at greatest risk. The shrinking left-hand side of the bell curve is a success we should celebrate and protect. The growing right-hand tail of severe obesity is the real tragedy, and it is something that needs to be addressed. People who fall into either category should be given any help that they need. Things will not be improved by wide-scale interventions that cut a few calories from everyone's diets. It will not help to shave a gram of sugar, or a teaspoon of fat from the national average consumption. Those changes mean nothing to someone whose BMI is 50, 60 or 70, and may actively harm someone whose BMI is 18.

If we really want to improve people's lives, then every effort must be made to understand the individual circumstances of each severely obese person. Given that there are so few successful interventions to help people in this category lose weight, in some cases the focus on shedding weight might have to end. Instead, we may have to start looking at ways of helping them live with their bodies. Perhaps, instead of impossible dreams of thinness, we should be trying to alleviate the damaging conditions that blight their lives. Maybe we could help people become healthier and more mobile at the weight that they are, rather than insisting on the improbable.

Of course for that to happen, we would have to accept that fat people have value. That they are capable of leading productive and significant lives, regardless of whether we find them pleasing to the eye. We would need to show compassion, and dispense with any notion that they can never 'be themselves' within their fat body. We would have to let go of our belief that they are fat through choice or moral degeneracy, and admit that they are just people. They probably do have faults, troubles and unfulfilled desires, but that is surely the case for all of us. To confront the problems that obesity causes, we need to stop believing in the 'otherness' of fat people, and start looking hard at ourselves.

7

WHY ARE WE SO AFRAID OF FAT?

There are worse things than being fat, and one of them is worrying about it all the time.

Peg Bracken

As I write, I am sitting at Heathrow Airport, massively early for a flight and taking a little time for an early lunch. In search of somewhere quiet, I have found myself near to one of the closed gates. There are only six people close by, with plenty of space between us all. I have food from one of the chain cafés: a toasted tortilla wrap with chicken and cheese, some crisps, a large chocolate cookie and a cup of tea. Not the healthiest choice I have ever made, but not the worst either.

A few seats down from me, another man is eating his lunch. He is extremely overweight. Although I cannot say for sure, in researching for this book I have been asking a lot of people their BMIs, and I would say that he is obese. From his size, I would imagine that he is in the Class II or III categories, probably Class II. He is eating his lunch, bought from the same café, and he has made choices that most people would consider healthier than mine. A chicken salad, a bottle of water and a small biscuit. As he opened the biscuit, two lean young men who were sitting across the other side of the seating area started nudging each other and laughing. Look at the funny fat man eating a biscuit, they whispered unsubtly. They had not reacted in the slightest when I was eating my lunch, even though I had gone for the larger cookie, and spilt quite a lot of melted cheese on myself while eating the wrap.

Their reaction is not unusual. The fat man clearly noticed their reaction, sighed to himself and seemed to shrink a little. Yes, he looked hurt, but he had seen it all before. I was annoyed, but did nothing. There were two of them, they were both bigger than me, and I was embarrassingly covered in cheese.

As a normal weight person, with no outwardly obvious health problems, my dietary choices face no scrutiny at all. Why should that be? There is no good evidence that a particular choice is worse for someone overweight than it is for me. I could be contributing to all manner of health problems through my diet, yet no one would care in the slightest.

Health advice, 'five-a-day', minimally processed foods, balance, moderation, deprivation, exclusion – they only apply to fat people. The world will not judge me if I avoid wholegrains, fibre, fruit or vegetables. It will not judge me if I choose to do no exercise. It will not sneer if I am seen in the queue at McDonald's, or snigger when I supersize my meal. Yet any risks associated with these things are equally likely to affect me.

Fat people are judged by different standards. If they choose a salad, people will point, laugh and say they are shutting the stable door after the horse has bolted. If they choose a pie, they have eaten all the pies. Every food choice will be judged. Even when it comes to moral issues such as veganism or sustainability, fat people get a rougher ride. The animal rights group PETA recently produced a series of vicious fat-shaming advertisements positioned on the back of buses in the UK,* which its spokesperson claimed were 'light-hearted'.[1] Reports citing obesity as a cause of global warming add nicely to the moral panic surrounding the 'epidemic'.[2] By being fat, you are not only destroying yourself, but you are slaughtering animals and fucking up the planet. Just in case you didn't already feel bad enough.

GLASS HOUSES

As I sit in the airport, I feel anger at the two men, and compassion for the man they are laughing at. But before I start casting stones, I probably need to confront how inclined I am to do similar things. As

---

* If you aren't reading this in Britain, you may not know that fat people are sometimes insulted by saying they look 'like the back end of a bus'. No, it isn't funny here, either.

someone obsessed with other people's shopping choices, I spend a good deal of time surreptitiously staring at people's trolleys in supermarkets (if you ever shop at Morrisons in Retford, I apologise if you have caught me doing this). This is partly because I work in the food industry and love getting free insights into what people buy and why, but it is also because I am extremely nosey. The problem is, if I see a fat person, or a family of fat people, my nosiness comes with judgements that I would never make if they were of normal weight. I assume that they are fat because of the contents of their trolley, despite knowing how illogical that is. It is the same when I see a fat stranger order something in a restaurant. I know how stupid and unpleasant it is to cast judgement upon them, but I cannot help myself. At no point do I think that someone is thin because of the food choices they make. I don't see someone with a perfect physique and think I will buy or order exactly the same as them, and somehow be transformed into their model of physical perfection. So why is it I feel so differently when it comes to fat?

Nichola Ludlam-Raine is a registered dietitian who supports obese people in weight management groups, and patients undergoing weight loss surgery. She spends time with some of the most severely obese people in our society, the very far right of the bell curve, sometimes with BMIs over 60. She told me that many of her patients do not eat at all during the daytime, choosing only to do so in the evening. They do this so that they only eat when they are alone, in order to avoid the scorn and judgement of others. Eating at this point of extreme hunger is likely to lead them to choose calorie dense foods, and to promote bingeing and excessive eating. Although it often means they miss entire meals, most will end up eating more calories in the long run.

Fat people are routinely abused and accosted in the street and have insults yelled at them as they go about their day. If they exercise, they are publicly insulted. If they dance, they are openly mocked. In researching for this book I asked people to contact me with their stories. I have included a few in the text, but received hundreds more. They detailed childhood bullying, physical abuse and shameless workplace discrimination. I heard stories of people driven to the point of a mental health crisis because of a constant stream of insults from strangers, friends and even family members. Lives lived in torment because society finds the way they look unacceptable, and approves their abusers.

In 2015, a group of men even started anonymously dispensing cards to overweight people on the London Underground. They would bravely target lone young women, handing over the following message:

> Overweight Haters Ltd – Our organisation hates and resents fat people. We object to the enormous amount of food resource you consume while half the world starves. We disapprove of your wasting NHS money to treat your selfish greed. And we do not understand why you fail to grasp that by eating less you will be better off, slimmer, happy and find a partner who is not a perverted chubby-lover, or even find a partner at all. We also object that the beatiful [sic] pig is used as an insult. You are not a pig. You are a fat, ugly human.

In the *Daily Mail*, Katie Hopkins was later given a platform to defend the man handing out these cards, saying that 'perhaps he was at his wits end with inaction on this issue'. Presumably she also despaired at his inability to spell 'beautiful', although I imagine she blamed this on immigrants taking up places in schools. I guess that is also why she failed to use an apostrophe when referring to the end of someone's wits.

I am sure that most people reading will agree that both fat-shaming-cowardly-tube-man and his number one supporter Katie Hopkins are worthy of rebuke. But it seems unlikely that if someone had been handing out cards targeting another minority, a national newspaper would have dared allow Hopkins, or anyone else, to voice their support.

BEDSIDE MANNER

We have already talked about the health risks of stigmatisation, and it is easy to appreciate how a lifetime of bullying and shame could take a toll on people. But stigma and prejudice may well be having a more direct and immediate impact on the health of fat people. Studies of healthcare professionals show a widespread belief that obesity is a problem of will-power, and is solely down to an individual's poor choices.[3] This is despite years of research indicating the massive influence of genetic and hormonal factors, and the well-known ineffectiveness of most weight control interventions. Widespread prejudice has been observed in doctors, nurses, psychologists and dietitians,[4] all of whom have a crucial role in the treatment and care of obese patients.

These studies also showed how doctors often view obese patients as annoying and ill-disciplined, and have less desire to help someone with a higher BMI. There is a widespread resentment of fat people within the medical profession, and a belief that obese patients are a waste of a doctor's valuable time. Perhaps even more worrying, there is a lack of willingness to refer obese patients for treatment, indicating that the entrenched stigmatised views of fatness can lead to a dereliction of their duty of care.

Nurses, too, often see obese patients as lazy and non-compliant, as do healthcare students, whose famously dark humour is often directed at fat people. Again, one of the main reasons for this is a widespread belief that obesity is a controllable condition, simply addressed by eating less and moving more.

This has led to a breakdown in communication between healthcare professionals and obese patients, with large numbers feeling too embarrassed to report problems, unwilling to subject themselves to an environment of judgement and stigmatisation. There is a huge health burden associated with the most severe cases of obesity, yet attitudes within the medical profession push away those in most need.

It is hardly surprising that this attitude exists within the medical profession, given that they are constantly being told that a fat apocalypse is threatening to destroy the healthcare system. If you are a doctor or nurse, working extraordinarily hard in a stretched, under-funded and crumbling system, the temptation to pour blame and scorn on a group of individuals that society insists is weak-willed and disgusting is perhaps hard to resist. They are certainly far from the only group that does this.

WORKPLACE

At the risk of agreeing with Katie Hopkins, many studies have shown that if you are fat, you are less likely to be successful in job interviews. Fat people are seen as less competent and with lower leadership potential. It is much harder for them to get to the interview stage if employers know about their body weight in advance. If they are lucky enough to be seen, they are much less likely to get the job. And even if they do, their starting salary is likely to be lower than an equivalent employee of normal weight.[5]

Once in the workplace, there will be fewer opportunities for promotion, lower pay, and a false perception that their performance is not as good as that of a normal weight employee. Most work longer hours, perhaps in an attempt to counter these perceptions. They are, not surprisingly, highly likely to experience bullying and stigmatisation. Sometimes this is vicious and cutting, but more often it is low level abuse, framed as humour, but designed to keep fat people firmly in their place.

Perhaps unsurprisingly, obese women are targeted far more than obese men, often judged unsuitable for work, particularly in customer-facing roles. Even in the twenty-first century, women are far more likely to be sexually objectified, with a much greater expectation that they should comply to cultural norms of beauty. Fat women defy these expectations, and so in order to hide how shallow we are, we project laziness and unprofessionalism instead. For all the progress in the workplace over the last fifty years, there is still an expectation that women should be pleasing to the eye.

Throughout the rise of female body acceptance, and the increasing prevalence of images depicting women of different sizes, female career success is still equated with not being fat. Popular images of successful doctors, lawyers, business women and professionals are almost exclusively thin.Thin women are more likely to be hailed as canny and motivated, whereas women with more curvaceous figures are assumed to be lazy and vacuous. Far too often, the difference in perception seems to be down to the size of the person's arse, not the evidence of whether they are hard working or talented. This should offend anyone who cares about fairness.

Healthcare professionals who are overweight themselves are routinely vilified, both by the public and the media. In 2016, the television hypnotist Steve Miller launched a campaign for overweight nurses to wear fat shaming badges to 'encourage' them to get thin. Although he received much criticism, his campaign also got a great deal of media coverage and support.

In a 2014 television debate, the cardiologist and diet book author Aseem Malhotra publicly attacked the dietitian Catherine Collins, suggesting that people should not take advice from her because she was overweight. Catherine is a legend within her profession, one of the most experienced, knowledgeable, intelligent and dedicated dietitians in the UK. She has years of clinical experience and an encyclopaedic knowledge of nutrition science. She contributed to my last book, and helped Ben Goldacre with his classic debunking text, *Bad Science*. Yet a medical doctor felt justified in questioning her capabilities in the face of the evidence.

This argument is played out again and again, with endless body shaming, and the equation of fat with capability, morality or intelligence. And it is particularly directed at women, perhaps because when a woman does not slavishly dedicate herself to attaining a form that men find sexually desirable, they are somehow considered a threat. And yes, I can confirm, men really are that shallow.

ANYTHING ELSE?

Fat people are more likely to be convicted of crimes by a jury.[6] [7] They are stigmatised in education, viewed as less capable and intelligent. Studies have even shown a high prevalence of weight prejudice among nutrition, public health and obesity researchers. In 2013, with all the sophistication of a hormonal thirteen year old, the evolutionary psychologist Dr Geoffrey Miller tweeted, 'Dear obese PhD Applicants. If you didn't have the willpower to stop eating carbs, you won't have the willpower to do a dissertation #truth'.

After a great deal of criticism, Miller deleted the tweet and apologised, yet the fact remains that a supposedly intelligent academic held such distorted and deeply entrenched views about fat. His views were especially concerning considering that Miller sat on his university's PhD review board. In the aftermath, a number of people produced data showing that Miller's view had no basis in fact, and that the real #truth was that there was no relation between BMI and the ability to complete a PhD, which will come as no surprise to anyone living in the real world.[8] Based on the constant abuse, fat people in academia are probably stronger and more determined than the rest of us.

Negative comments about fat people are more common and far more accepted than comments about race. A 2008 study showed that weight prejudice was significantly more engrained and accepted than bias against sexuality or religion.[9] It seems that weight is also one of the hardest prejudices to fight, with many of the techniques that are successful in countering other prejudices failing to work when it comes to fat. Attitudes towards fat people are robust, and exceptionally hard to change, and tragically they contribute to a great number of mental and physical health problems.

Fat prejudice has also, perhaps surprisingly, increased as the prevalence of obesity has risen, accelerating considerably in the last forty years or so. Many media commenters warn of the dangers of obesity becoming normalised, but we are in no danger of that happening. A 2015 study by Dr Stuart Flint, a psychologist at Leeds Beckett University, showed that it was most common in men, especially young men who exercised frequently. Females were found to have less explicit bias, and were more accepting and sympathetic towards larger people, perhaps reflecting greater societal expectations of caring and empathy among women. But interestingly, when implicit attitudes were studied, using techniques to investigate people's hidden biases, women showed exactly the same levels of bias as men. Although outwardly more compassionate, they were only hiding their prejudice below the surface.

Underlying all these biases is a common assumption that obesity is controllable, and that fat people are to blame for their condition. An acceptance of this belief is highly correlated to prejudice, something that is seen across all groups, even in the studies of healthcare professionals and obesity researchers.

Unsurprisingly, overweight and obese people are more inclined to accept the uncontrollable nature of body weight, presumably due to their own life experiences. But Dr Flint's study also threw up something a bit more interesting. Underweight people were also more likely to accept that weight is not just a matter of will, perhaps because they too are aware that it is sometimes impossible to achieve an aesthetic ideal through food choices alone.

Worryingly, weight stigma was generally more prevalent among younger people, despite so much progression in areas such as race, gender, sexuality, disability and religion. Perhaps this is because young people are naturally more image obsessed, concerned with physical attraction in a way that thankfully fades as middle age approaches. Or maybe it is the result of a technology-led generation that places increasing importance upon visual images, defining themselves by how they look on a selfie, constantly subjecting their posts to the scrutiny of the world. Maybe it is a symptom of the increasingly common use of electronic dating apps to select a partner, rather than the clumsy approaches in the pub that previous generations favoured. Could it be that having a million potential partners to choose from results in an obsession with early filtering, shutting out people of a certain shape or size, deeming them of less value than other categories? Perhaps the limited pool of potential partners available to previous generations meant that less attention was paid to superficial criteria. Or maybe it is just that young people have lived their whole lives with the dominant neoliberal narrative that people are fat because they are weak, and should only be accepted into society when they can be bothered to get thin.

Whatever it is, our bias comes from somewhere. And if it is to be fought, we need to understand exactly where.

THE NEWS

Here's what the UK press has to say about obesity.

> No one is arguing for 'cigarette smoker acceptance' because we know that fags cause ill health and smokers bring it on themselves. Being obese is the same thing, yet somehow criticism of being overweight is conflated with racism or sexism. You don't choose your race, gender or sexual orientation, but other than in the drastically rare cases where you have a metabolic disease, you do choose to be overweight.
>
> Nick Mitchell, personal trainer,
> seemingly allowed to advertise in the *Telegraph*

> Samantha Packham is Britain's youngest person to die from morbid obesity. At 20 years old she weighed 40 stone [254 kilograms or 560 pounds]. I met her parents and challenged them on their care for their daughter. They allowed her to take taxis to get more food. I believe they fed their daughter to death.
>
> Katie Hopkins, *Daily Mail*

Studies of news articles on obesity have found that they are almost universally negative in tone when it comes to descriptions of fat people. The vast majority focus on behavioural reasons for weight gain, especially lack of exercise and overconsumption of food. Precious little space is given to genetic and hormonal control, despite the emergence of this field in recent years. There is also a tendency for the media to portray obesity as a female problem, with greater disgust for fat female bodies a strong selling point. But women are also blamed for making their children fat. When bad parenting is blamed for obesity, we can usually read this as an attack on mothers, selfishly returning to the workplace, and refusing to spend hours slaving in the kitchen to put delicious home-cooked food on the table.

It is estimated that over 70 per cent of images accompanying obesity stories show fat people in a stigmatising way. Common approaches are the previously mentioned torso-only shot, fat people in ill-fitting clothes, fat people half undressed, fat people eating notoriously unhealthy foods, or any combination of these. Most of these shots pretend to be candidly capturing fat people in their everyday lives, but they are almost always contrived. Clothes are chosen not to fit, exposing rolls of flesh bulging over tight waistbands. Prop burgers or pies are placed in hands, styled to look as cheap and unhealthy as possible. Models are instructed to appear lazy, sad and aimless. There are frequently comparison shots against thin superiors, leading infinitely better and happier lives.

These images are absolutely pervasive. A year ago, I wrote an article about weight stigma for a well-known national publication, only to be shocked when it went to press accompanied by an image of a disembodied fat person in ill-fitting clothes, sitting on a park bench, burger in hand. Although I stood by the article, I was ashamed that it was shared in this way. The media constantly reinforces the idea that obesity is the fault of the individual and they are only fat because of their actions. A study of news media stories by Dr Stuart Flint in 2015 found that 'there is clear evidence of promotion of Protestant ethic values in newspaper portrayals of obesity, presenting obese people as immoral, slothful and gluttonous, and expressing the view that obesity is akin to deviant and immoral behaviours.'

Even when bullying of fat people is covered, often the articles carry little criticism of the bullying behaviours, and sometimes even frame the bullying as a motivating factor in helping someone lose weight. Of course, they never consider the possibility that the stress from the bullying caused weight gain, and only when the bullying had abated was weight loss possible. The same newspapers that cover bullying will often carry jokes and cartoons ridiculing fat people, and will probably

be responsible for publishing exactly the sort of blame and shame articles that contribute to it in the first place.

Dr Stuart Flint told me that 'these attitudes are formed from
consistent media messages. There is very little in the way of detriment
penalty for holding or expressing these beliefs. In fact, in many situations it can be socially beneficial. When Rafa Benitez was manager of
Liverpool, there was a popular "Fat Spanish Waiter" chant that went
around, and that was an accepted and socially bonding form of abuse.'
He goes on to suggest that it is not just the media than enforces these
attitudes. 'A lot can come from politicians too. David Cameron was
always very keen to frame obesity as a behaviour, and to equate it to
drinking, smoking or drugs. He positioned this as speaking out against
moral neutrality, but it is equally stigmatising. Sadly, this is all very
effective. Conditioning interventions that can work on reducing racial
bias are not as effective when it comes to weight bias.'

Despite signing commitments to avoid stereotyping people regarding their 'race, gender, age, religion, ethnicity, geography, sexual orien-
tation, disability, physical appearance or social status' (Society of
Professional Journalists 2010), journalists frequently flout these guidelines, and very few of them are ever reprimanded.

THE DRUG OF THE NATION

As well as the news, television also plays a very important part. Obese
and overweight characters on television have fewer friends and romantic
relationships. And as the landscape of television has changed, so have
depictions of fat people. In the early 2000s, as channel numbers
increased, and a demand for accessible, compelling reality television
formats grew, fat quickly became a form of entertainment. Weight loss
shows such as *You Are What You Eat*, *The Biggest Loser* or *Supersize vs
Superskinny* grew in popularity, showing how a fat citizen could be neatly
transformed into an ideal one, replete with a brand new spray tan and
makeover. Quickly, reality television became the main place that people
saw depictions of fat bodies. And with a strange circularity, reality television viewing became synonymous with sloth, sin and fatness. People sat
at home on their increasingly fat arses, watching reality television
programmes full of fat people, who were only fat because they spent all
their time sat on their fat arses watching reality television.

The shows would start with the lifestyles of obese people forensically invaded for our viewing pleasure. Experts audited them, picked apart their flaws, explained the reasons why they had failed, and provided a conveniently simple plan for transformation. The food consumed in a week of sinful existence would be laid out on a table, dripping with saturated fats and heart clogging poisons. Wheelbarrows full of sugar representing their annual sweet indulgence would be thrust in front of the fat sinner, just to make sure they felt truly ashamed. Shocked and tearful close-ups would follow, the emotion of realising what a terrible citizen they had been clear for all to see.

Then the transformation stage, full of drama, pain, defeat and triumph. The fat protagonist is guided along a traumatic path to health by thin, attractive escorts, graciously allowing them the chance of a better life. Soon, at a time predetermined by production schedules, a thin nirvana is reached. The new smaller citizen is paraded around, pleasingly taking up far less space, happier, healthier and on the path to a better life. Obviously, given the short timescales and the huge difficulties of weight loss, the transformation often involves new clothes, high heels, flattering camera angles and a makeover. But this hardly matters, because the errant fat person had been tamed. One day, they might even be loved.

These programmes are the ultimate expression of neoliberal values, played out as trashy, predictable dramas. Be a good fat person. Act like the thin people tell you to, starve yourself for our amusement, let us laugh at your pathetic attempts to exercise, and we might allow you a piece of our world. Work hard and we will give you some of our shiny things. The nice clothes, the stylists, the spray tan. Combined with a rising moral panic about the obesity epidemic, weight loss television was perfect reality fare. Neat three-act dramas, with tension, jeopardy and a simple common-sense resolution.

Jayne Raisborough is a professor of sociology at Leeds Beckett University. She has studied the way obese people are portrayed in the media, with a particular interest in these television programmes. She told me:

I started thinking about what instructions we are being given. What exactly are we being taught by these programmes? What you soon realise is that there is no advice being given out at all. In a programme about weight loss, you certainly don't learn what you think you are going to learn. All you find out is that we have this body intolerance, and if people do not fit the body norm, they need to be changed. You learn to do a living autopsy on fat people, and you start to relate diseases to character traits.

Jayne describes this as a move away from absolute humiliation to a more benign, but equally damaging form of stigmatisation. In helping fat people 'cure themselves' of their bodies, a thin elite appears to be tackling the problems they face. In the same way that seemingly benevolent forms of sexism encourage and reward female compliance, benevolent fatism rewards 'good' fat people, accepting them as proto-citizens, so making them strive to comply with their new masters' wishes. No wonder I always hated Gillian McKeith.

A 2015 study found that television was the primary source of advice for weight loss in the US, and it can be imagined that the same is true for many other countries.[10] But even bearing that in mind, it's still not as bad as . . .

FASHION

No one wants to see curvy women . . . You've got fat mothers with their bags of chips sitting in front of the television and saying that thin models are ugly.

Karl Lagerfeld (a man with a ponytail)

If there is any group that should bear maximum responsibility for our fear of fat, it is the fashion industry. The idolisation of thinness has strong roots in the fashion world, and the continued pressure on models to be dangerously thin is a lasting shame that it refuses to address. Fashion designers have long demanded a parade of shape-less, sexless, joyless female models because anything else would risk showing up their inability to design anything to fit a variety of forms. The thinner the model, the simpler the engineering, the less chance that inconvenient feminine shapes will interfere with the oh-so-precious art.

Certainly some models are naturally thin, but there is no doubting the pressure placed on often very young women to conform to near impossible standards. They are treated as little more than coat hanger wire, not daring to take on the slightest shape, lest they deform the exquisite work of the designer. New collections are sent to fashion magazines in ridiculously small sizes, pressuring editors to find the thinnest models available, so increasing the constant drive for emaciation in order to get prestigious work. It is a vicious cycle of destruction that has ruined many young lives, both in and out of the industry.

Models are starved into illness and disorder, and so are the women who try to emulate them. Thinness comes at a cost, both mentally and physically, with heart conditions and anorexia commonplace among top catwalk models. Model scouts have been known to talent spot outside of eating disorder clinics,[11] keen to find women capable of the dangerous restriction required to get work. And all because a few designers are too incompetent to create clothes for women, and too callous to care for the lives they are destroying.

Out of all this comes a bizarre and unattainable vision of female beauty. It is a sexless beauty of control and denial. Even for the small number of genetically appropriate models, it is a beauty that is nearly impossible to achieve, requiring consistent self-harm for the length of their career. It is a career that carries with it a severe risk of early death. And for everyone else, it places beauty ideals out of reach, leaving women constantly chasing an impossible dream.

Many of the thinnest models are likely to have a BMI in the underweight category. For a few, this will be their natural weight, and this can be a hard thing to change. Katherine Flegal's work showed that the mortality risks of being underweight are many times higher than those associated with being fat. When the fashion world has so much influence, can it really be right that they idolise and celebrate a body shape that carries such a high risk of death? And why is it that every time a plus-size model takes to the catwalk, some deluded commenter claims that we are sending out damaging messages? The 'plus-size' catwalk models probably have the lowest weight-related mortality risk of anyone in the room. Is seeing a normal, healthy woman wearing nice clothes really going to ruin people's lives?

I do not understand the fashion industry, and suspect I never will, but I do see the power it has to set expectations of physical beauty. It seems to me that a rank incompetence to make clothes fit blights this industry. I cannot understand how the world can value designers who consistently fail to make clothes that people can wear, when this is their only job. I see no value or beauty in this work, and certainly nothing that could justify the vandalising of countless young lives.

Models are promised a life of privilege if only they can become thin enough. To do this, they must starve. But starvation makes them physically weak, and tired. Elite men surround themselves with these thin, broken women, broken by a slavish desire to fit into a poxy fucking dress. Fashion profits by creating and reinforcing fear. I might not understand fashion, but I know that it is insidious and vile. Whatever worth there is in the art the industry produces, it does not justify the terrible price that people pay for it.

WHAT ARE WE THINKING WITH?

What exactly is it about fat that we hate and fear so much? And what is it that drives our intense fascination? Although societal attitudes, the media and the fashion industries certainly play a major part, there are perhaps deeper reasons for our lack of acceptance. It is likely that sexual attractiveness plays a part, and may have deep evolutionary underpinnings. Certainly, hip to waist ratio is a robust indicator of perceived attractiveness in women, and with so much value placed on measures that men find sexually desirable, perhaps a rejection of fatness is connected to some feeling that fat women do not fulfil a useful role in society if men don't want to have sex with them.

I asked the Harvard psychology professor Steven Pinker why he felt there is such a connection between body size and perceived morality:

> I suspect it's a combination of two aesthetic reactions. One is our positive response to sexually attractive individuals, and to healthy, non-deformed individuals, and the accompanying negative response to unattractive and unhealthy ones. Sexual attractiveness generally tracks superficial cues to fertility (e.g., youth in women, bodies of the right shape for each sex) together with health and bodily integrity. All are a mixture of universal tastes and receptiveness to the statistical distribution of the population one finds oneself in. This in turn is distorted by the 'halo effect', which is part of our tendency to moralize our gut reactions: things that we sense as good and desirable are also thought of as moral, and those that make us queasy blur into the immoral.

This is undoubtedly true, but it does raise the question, why is it that as our statistical distribution of weight has changed, our definition of what is considered attractive has not? If fact, in many cases it appears to have moved further down, with leaner women being seen as more desirable as average body weights have risen.

Exactly where this comes from is complex and uncertain, but given that it has changed much over time, it seems doubtful that it is innate. Data from Google search terms for pornographic images seems to indicate that many men find larger women more attractive than they are willing to admit publicly. When comparisons are made to their preferences on dating sites, where they will often favour women of lower weights, it seems that many are choosing partners based more on a pressure to conform to the expectations of society than on what they actually desire.[12]

It is likely that we learn some of our prejudices and expectations from the influence of our immediate family, passed down from parent to child. Kimberley Wilson is a chartered psychologist and food writer, and former Chair of the British Psychological Society's Training Committee in Counselling Psychology. She has many years of experience counselling people with weight and eating concerns, and told me:

> Parents can see children as an extension of themselves. If a child is considered 'less-than-perfect' some parents are driven to change them, and this can reinforce the value of being 'slim and attractive'. This is pernicious and persistent. Mothers of a certain class will send their young daughters to nutritionists if they gain weight, and consistently teach them to value the aesthetic. I have known mothers that pay their children not to eat, or tell their healthy weight daughters 'you will feel better about yourself if you skip dinner'. Children equate how valuable they are to their parent with their value to society, and so people who do not comply to this aesthetic are seen as being of lower value. This can cause many different problems. It may create people with stigmatised attitudes to weight, but it can also end with rebellion, either against deprivation, or against parents' love being conditional upon appearance.

Although women are often the target of weight stigma based on a failure to be sexually attractive, for men there might be other factors involved. Fatness is sometimes associated with bodies becoming less masculine, growing lumps, curves, breasts and softness associated with femininity. In many ways, fat men lose their male status, and at a time when male roles are in crisis, this apparently wilful destruction of their own masculinity is viewed with disgust. If we all get fat, women will take over the world, condemning us to subservience or, God forbid, equality.

In the 1950s, a Canadian campaign to fight rising body weights in men centred on the dangers of an out of shape population being unable to defend itself in a potential Cold War conflict.[13] 'Chubby Hubbies', as they were termed, would not be able to rebuild Canadian society after a nuclear attack, so wives were cajoled into getting them into better shape. This trope of masculinity still persists, and all men secretly imagine that they might one day be called into action.

Anyone who has seen the sort of fat shaming that emanates from the male fitness community on social media will recognise this. There is good evidence that the sort of obsessive body sculpting seen in these communities is far from healthy, with a culture of dangerous drug use at its heart. Strength, not health, is idolised. Fat men are shamed, because their slothful bodies would be useless in a fight. Never mind that they are grown-ups in a civilised society and violence is the recourse of the stupid.

Kimberley Wilson also believes something else might be going on: 'There is also the possibility that it might be something to do with "Reaction Formation" . . . In a society where we are all compelled to comply, outliers are admired and yet despised. These are people testing the boundaries of what is respectable. The idea of a happy fat person causes discomfort because they have deviated from the norm. Sometimes we despise the things we secretly admire.'

It is quite tragic that many men find themselves stuck in a world of deprivation and pain, just because they fear their fragile masculinity would not survive a few pounds of excess weight.

WHAT TO DO?

Addressing this prejudice forms no part of government obesity strategies.[14] Currently, the WHO suggests that we should try to help obese children and adults develop 'mental health resilience', instead of making the prejudice unacceptable. Our plan is no more complicated than 'Man up, fat people'.

The most effective method for combatting weight prejudice involves showing people the evidence for how innate and uncontrollable body weight is.[15] But despite the robust science that underlies the genetic and hormonal models of weight regulation, 'eat less, move more' still dominates the media, public health strategy, healthcare interventions and government policy.

It is a matter of choice whether or not we stigmatise people because of their physical appearance. We can just decide not to if the will is strong enough, as evidenced by the extraordinary, although sadly incomplete, progress in our attitudes towards sex, sexuality, disability and race.

Shaming fat people does not help them. It makes their lives harder. It makes their health worsen. The evidence for this is just as strong as for the harms of obesity. It is likely a huge confounding factor in the health problems commonly associated with being fat. And even if it didn't cause such harm, it is just a shitty thing to do.

This is a plea to all of us to stop our subtle jibes and jokes. To put an end to our judgemental looks in restaurants and supermarkets. To look past the weight of an interview candidate, a shop assistant, a patient or a friend. To listen to people's concerns and see value beyond their size. To stop believing that someone's worth is defined by how they look. And to let go of the idea that being fat is anything to do with poor choices or a lack of character. If we are really as keen as we seem to be on self-improvement, let's try to be better people.

PART II

WHY ARE WE SO FAT?

For every complex question there is an answer that is clear, simple and wrong.

H. L. Mencken

8

IS IT BECAUSE OF OUR GENES?

There is a universal response to not having enough food, but no universal response to having too much.

Dr Giles Yeo, President of the British Society for Neuroendocrinology

I am often told by well-meaning thin people on social media that no one is born obese. As far as I am aware this is literally true, although calculating the BMIs of newborns would be a ridiculous waste of time. It is an accusation frequently levelled by those who perpetuate stigma against fat people, and they go on to explain that, unlike race or sex, obesity is something controlled by our behaviour, not defined by our genes.

It's ignorant and deluded to argue that you should be allowed to bully someone for a characteristic that's not defined at birth. That would imply that prejudice against religion, sexuality and many types of disability is absolutely fine, as no one is 'born' any of these things. Even when it comes to racial prejudice, this has far more to do with cultural context than the colour of someone's skin.

We are all the product of an interaction between our genes and our environment. We are not born speaking or walking or having sex or able to digest solid food. Any claim that 'no one is born obese' fails to understand this basic truth, that we are not really 'born' anything, other than human. Character is affected by our environment, and certainly not linked to our physical appearance (unless we are bullied for it).

One thing we are definitely not born with is prejudice. A newborn child cares nothing for skin colour, body size, religion, sexuality or disability. Prejudices are taught to us by the society in which we live. However innate and logical they feel, our hate towards specific groups of people is learned, and can be countered with education. A scientific understanding of how unchangeable genetic factors shape who we are can be extremely important when it comes to shaping our cultural attitudes.

The genetics of race shows exactly how few intrinsic differences there are between people with differently coloured skin, revealing the rank stupidity of all racist beliefs. Education is key in the countering of prejudice, and genetics can be at the heart of this. Adam Rutherford's book *A Brief History of Everyone Who Ever Lived* eloquently explains how tiny genetic variations produce the physical differences we associate with race, and how this in turn can have a profound effect on the way in which we treat people. And as we shall discover in this chapter, what is true regarding race is very much echoed when it comes to weight.

But hang on, the doubters will say. Weight is surely determined by how we interact with our environment. Energy in, energy out. Eat less, move more. Even if you are genetically susceptible to weight gain, you have more control over your weight than your skin colour. Yet the more we learn about the genetics of obesity, the harder it becomes to agree.

SKINNY GENES

In Chapter 1, we discussed studies that showed body weight to be highly heritable, leading researchers to selectively breed fat mice in order to investigate the genetic causes. We also discussed the concept of a set point weight, and the more nuanced idea of a settling point, influenced by genes and the environment. Although set and settling point weights are a little contentious, the idea that body weight is strongly genetically determined is beyond dispute. Numerous twin studies have shown that identical twins, sharing all of their genes and living in the same environment, are far more likely to be of a similar body weight than non-identical twins, who share the same environment yet only half of their genes.

Obviously, identical twins share many characteristics, particularly physical ones, and several of these have a strong genetic basis. In identical twins, we would expect little variation in childhood hair colour for instance, and so this is considered to be nearly 100 per cent heritable. Height also has a strong genetic component, but can be influenced by environmental factors such as diet, lifestyle and illness, and so is not as heritable as hair colour. Twin studies have shown the heritability of height to be around 80 per cent.

The heritable component of body weight has been studied a great deal, and although there is some variability, it is thought to be around 70 per cent heritable in adults. In children this figure is even higher, around 80 per cent, with the effect of family and environment having virtually no influence at all.[1] Even separated twins, adopted apart from each other at birth and raised in different environments, have as much similarity when it comes to weight as twins raised together.[2]

These figures make body weight a good deal more heritable than many other factors that we generally accept are strongly under the influence of our genes. These include conditions such as schizophrenia, breast cancer, hypertension, alcoholism and heart disease. But it also puts obesity above behaviourally defined characteristics, such as sexuality,[3] that only the cruellest would single out.

Strangely perhaps, over the years, as environments have changed and food has become more readily available, the heritable component of obesity has actually grown. It appears that when food is in short supply, the main determinant of how much energy you store is the environment, but as food becomes more available, previously hidden tendencies to store fat are revealed.

Although in many cases the genetic causes are complex, around 15 per cent of people with BMIs over 40 are thought to have a single, identified gene defect.[4] As well as the rare leptin-deficient children discussed earlier, many have mutations in the genes that code for leptin receptors or other brain regions linked to the hormonal control of appetite. Obesity, it turns out, is one of the most heritable characteristics ever studied.

BUT, BUT . . . THERE'S AN EPIDEMIC

Despite obesity being irrefutably based in a person's genetic inheritance, publicly expressing this view causes consternation. There is, as we are forever being told, an epidemic of obesity, and it has happened recently. Even if we discount the epidemic narrative completely, there is good evidence that body weight has been increasing steadily for the past century, and it cannot be denied that the population distribution of BMI has changed since the 1970s. This seems at odds with body weight being under strict genetic control. After all, genes are relatively fixed, and are not likely to have altered much within the past fifty, or even hundred, years. Modern society places little or no selective pressure on populations, and so the potential for any evolutionary change is virtually non-existent within such a short timeframe. Given that our genes are unlikely to have changed, it seems like the obesity epidemic has been caused by environmental factors.

There is of course some truth in this, but height has also been steadily increasing since the beginning of the twentieth century, and there is no suggesting that this is not strongly controlled by genetics. Also, there is some tantalising evidence that recent alterations to our genetic make-up might explain some of the changes we have seen of late, despite how unbelievable that might sound. We shall come to that later in the chapter, but first it is worth having a look at what is currently known about how and why our genes make us fat.

WHEN YOUR GENES DON'T FIT

Dr Giles Yeo studies how the brain controls our body weight and the genetics of obesity. He told me:

Our genes haven't changed recently, but the environment has. Genetic changes can either be positively selected for, negatively selected for, or neutral. Sometimes, genetic changes can be neutral or positive in one environment, but if the environment alters, they can become negative. At the moment, for instance, a chef and a geneticist can both make a living, but if there is an apocalypse, suddenly being a geneticist becomes neutral, but being a chef is positively selected for, as people need to eat. Or take skin colour. Dark skin protects from the negative effects of strong sunlight, but when people moved to colder climates, their skins gradually became lighter as a positive adaptation. When pasty Northern Europeans suddenly started moving to places like Australia, they developed a much higher incidence of skin cancer. Their once positive genes suddenly became toxic. What is important is the way that genes interact with the environment.

Once we know that a particular gene has an influence, we can start to gather information about how it might work. For example, the researchers who discovered leptin also found that the gene was expressed in fat cells, which provided the first indications that stored fat was involved in the control of appetite. This gave us a potential mechanism to explain how a set point weight might be regulated in the body, opening up new avenues for research.

Although single gene alterations such as the leptin mutation are rare, there is increasing interest in how a large number of different genes might act on the weights of most people. Giles Yeo explains:

Aside from genetic diseases such as the inability to produce leptin, we have learned from genome studies that there is a big genetic influence on body weight and body shape. We also know that these two are genetically distinct. The genes that control body shape, particularly waist to hip ratio, are involved with fat biology and are independent of BMI. There are separate genes that influence the BMI, and these are largely enriched* in the brain and central nervous system. There are over 100 of these, and they are involved in things like controlling appetite, hunger and reward behaviours. Obese people tend to be genetically driven to eat more food, and to be hungrier. While there is a choice over what decisions they make, they have no choice over their hunger. Although people can choose not to do something repeatedly, it is very difficult to say no to hunger all the time.

Many of the individual genes being investigated by scientists like Dr Yeo have a very small effect on overall body weight. Population studies show that people with a particular variant of a typical gene might have an average BMI difference accounting for only about 50 grams of weight. Others can have a greater effect. Certain variants of the FTO

---

* How much a gene is enriched refers to how much it is expressed in a certain situation. Essentially it is the degree to which that gene is 'turned on'. Most of the genes related to obesity are enriched in and around the hypothalamus.

gene,* for instance, can account for differences of around 3 kilograms, making it a hot area of study. Like most of the genes of interest, FTO is thought to be expressed in the hypothalamus, and seems to be involved in the regulation of food intake.

Although individually a few grams here and a couple of kilograms there might seem trivial, when added together, the prevalence of several of these alterations can be quite significant. The vast majority have effect sizes less than about 150 grams, but there are many of these, and in combination they can potentially have an important effect.

This field of study has yielded some interesting results and insights. More than anything, it seems to indicate that most of the variance in BMI comes from genes expressed in the brain, particularly the basal control mechanisms, which are primitive parts of the brain common to all vertebrates. This indicates that behaviour largely controls our weight, rather than anything more fundamental to our digestive systems. Of course, this doesn't mean obesity is controllable – just that the brain is probably the place to look for answers.

But there is a problem. When the total amount of weight influenced by the 100 or so known genes is totalled up, it only accounts for about 5 per cent of the variance known to be heritable. In all the years of study, with all the resources thrown at increasing our understanding, and despite having fully sequenced the human genome, we only know where one twentieth of the heritability is coming from. This has prompted many people to ask if we are looking in the right place, or if the field of study is somehow flawed. In reality, however, there is probably a simpler explanation. Giles Yeo again:

> We all have thousands of unique genetic mutations, and it is only the most common ones that are studied. At the moment, technology is used to identify the 100 known genes. Five per cent of variation can be accounted for by these known variations, but if you wanted to get the full picture you would really need to be looking for the unknown unknowns. To really understand what is going on, you would probably need to do whole genome sequencing for a large number of people, but that would cost around £2,000 per person, and we just can't afford it.

---

* There are too many genes to give them good names.

In reality, the missing heritability will almost certainly be discovered as techniques to sequence the whole genome become faster and cheaper. Current techniques are limited because they only search for a small number of specific changes to certain genes. It is highly likely that other alterations in and around those genes also have powerful effects yet to be discovered, but until techniques improve, it is not possible to look everywhere.

One of the key insights of modern genetics is that there is not a specific gene for everything, but instead a number of complex interactions. As more sophisticated technology becomes available, scientists like Dr Yeo will doubtless get closer to a complete picture of these interactions, and a better understanding of how our genetic code influences our body weight.

WHY WE EAT

Dr Clare Llewelyn is a behavioural obesity researcher at University College London. Her work investigates the genetic basis for the two known systems of appetite control. One is satiety sensitivity – how easily you fill up once you have started eating. The other is how responsive you are to food cues – how much you will be driven to eat something that looks, smells or sounds delicious. She told me:

> The first system is an ancient part of the brain, the second is a more modern reward system. The two are independent, but they do talk to each other. For instance, you can feel completely full after a meal, but still somehow find room for dessert if it looks delicious enough. We have used twin studies to look at the genetic basis for this. We have found that there are massive differences in babies from just a few weeks old. Most researchers believe that all babies are born with an ability to self-regulate, but we have found that this is not the whole story. Babies vary from day one as to how much they eat, how demanding they are and how easily satiated they are. In infancy satiation is 72 per cent heritable, and food responsiveness is 59 per cent heritable. Some babies are genetically prone to overeating, some prone to not feeding well.

Given what we have learned so far, it might not follow that overeating always leads to weight gain. But Dr Llewelyn has kept going with the study. The twins she has been studying are now approaching ten years old, and it seems that there might be a connection.

> Low satiety sensitivity and food responsiveness are character-ised by distinctive patterns of overeating. We are finding that children with blunted satiety sensitivity tend to eat a bit more, every time they eat (they need a bit more food to feel full and satisfied), while children who are more responsive to food tend to eat more often throughout the day. Food responsive children eat an average of three extra meals per week. We found no difference in the energy density of diets, and they generally have the same amount of carbs, fat and protein. There are actu-ally no vast differences, just small changes every time they eat.

Obviously this begs the question, what should you do with a child who is inclined to eat to excess? We have seen the potential for harm caused by dieting and food restriction in young people, but when a child has an innate tendency to eat too much food, should they really be allowed to do so indiscriminately?

It is important to note that until the children being studied have become adults, the long-term implications of these differences are not known. But for now, Dr Llewelyn suggests that 'it is important to understand what sort of eater your child is. A hungry baby is likely to be a hungry toddler and a hungry child. If a child overeats certain types of food, maybe be more vigilant about portion sizes when serving that particular thing.'

One thing is clear. Genes determine much about our appetite and our weight. But why exactly is it that a small collection of genes that affect primitive parts of our brain control the amount of fat that we store in our bodies? Why do some of us seem genetically predisposed to gain weight, while others stay effortlessly thin?

EATING AND GETTING EATEN

We have already discussed set point theory and the contrasting settling point. Although they sound similar, the two theories represent a split in the obesity science community, between those who see it as a fixed result of genes and those who see it as a variable result of behaviour. The reality is probably that neither viewpoint is fully correct. Understanding from animal models, and recent population-level changes, shows that set points are certainly not fixed. But similarly, people's inability to lose weight, and the strong genetic determinants, shows that fat is not fully within our conscious control.

Aside from set point and settling point, there are a few other models. For instance, the catchily titled 'dual intervention point' suggests genetically determined upper and lower boundaries for our fat stores, with the body only intervening if we should fall outside.[5] This makes a good deal more sense than a single defined weight that we cannot stray from. Professor Traci Mann from the Healthy Eating Laboratory at the University of Minnesota has suggested that the best weight management strategy is for people to find their lowest weight they can easily maintain, perhaps represented by the lower boundary in the dual intervention point model. This might not be exactly the weight we want, or that the medical community deems ideal, but it could be the healthiest for that individual.

Whichever model is used, there are some important questions to be asked about set point weight. Firstly, why exactly do humans have one in the first place? And perhaps more importantly, why might the ability to control it be different among individuals? Why is it that I have remained thin throughout my life, but others are locked in a constant struggle?

Professor Jeffrey Friedman has contributed a huge amount to this field. He led the research team that discovered the hormone leptin, developing much of the insight into how it functions in the body. In recent years, he has investigated the evolutionary basis for the differences in our ability to maintain a constant weight. He told me:

The existence of a set point suggests that adaptive evolution can select for differences in the amount of fat that a population carries. Indeed, there is good evidence that the amount of fat that is stored is under evolutionary pressure. For example, having large amounts in storage would enhance survival during periods of famine. But there is also evolutionary pressure against being too heavy because reduced mobility can lead to an increased risk of predation as well as problems with thermoregulation especially after vigorous activity. Over long periods of time, within a species, evolution has had to balance the risk of starving against the risk of predation.

So over the years, evolution has balanced the risk of starving and the risk of becoming another animal's meal. Friedman believes that our genes may have evolved in response to different environments. On a Pacific island with fewer predators but a greater risk of famine, evolution would select for fat storage, but in a place where both food and predators were abundant, it would be better to be lean and quick off the mark.

There is clearly wide genetic variation in how we respond to our food environment, and these evolutionary pressures provide a pretty good explanation as to why this might be the case. Certainly, Pacific island populations that lived for many years in harsh conditions proved highly susceptible to obesity once food became more plentiful.[6] So it may be that human populations carry the genetic legacy of different environments, giving us all different set points. Although many would argue that we should be able to override the effect of our genes once we know that starvation or predators no longer threaten us, this ignores the great power our bodies hold over our weight. Jeffrey Friedman says:

> In my view, basic drives almost always win over cognitive drives, especially over longer periods of time such as months or years. When lecturing on this topic, to make this point more concrete I offer anyone in the audience $5 million if they can stop breathing until the end of the lecture. No one takes me up on it. However, because the basic drive to eat plays out over a longer timeframe, people assume that this is because one's willpower has lapsed rather than that a basic drive has asserted itself.

When I ask him why people's weight has been increasing, and why the distribution has skewed to the right in recent years, his answer is disarmingly simple: 'Those people may not have been getting free access to calories in the past.' That idea has fascinating implications for the link between poverty and obesity.

Whether variable or fixed, the biological mechanisms by which set point weight is determined are poorly understood. But this lack of understanding is only because it is such a vastly complex area. There are likely to be hundreds of genes involved, from the hormonal regulation systems we have already discussed, through to genes affecting food absorption, metabolism, energy expenditure and body composition. Our knowledge of genes contributing to factors such as height, inflammatory bowel disease and a number of neurological disorders is far more complete. But obesity is more complex than any of these.

BECAUSE EPIGENETICS

Something else that might have an effect on obesity is epigenetics, which is the study of how different genes get turned on and off in the real world. This is an interesting field of study, although it does tend to attract more than its share of loons and cranks. There are ironic awards in academia for the misuse of epigenetics in scientific literature, and it is often mangled and contorted into vaguely sciencey sounding woo.

Here is an example from old Angry Chef favourite Natasha Campbell-McBride, who sells fad diets to vulnerable people. On the FAQ section of her website, in reply to a parent whose child has been diagnosed with a specific gene mutation causing a rare metabolic disorder, she cheerfully replies: 'I would recommend that you read about the new science of epigenetics. Our genes are not our destiny at all! Instead it is our diet and the environment (inside the body and outside) that pre-determine which genes get expressed and which genes will stay dormant.'

'Alternative' medicine types like to use epigenetics as 'evidence' that your medical destiny is not controlled by your DNA. Even more tantalisingly, they often suggest that you can somehow reprogram your genetic code in such a way that will allow you to pass on those changes to your offspring. This is usually used to motivate compliance with whatever bullshit they are hawking this week, but it is also frequently used to denigrate the choices of others. The suggestion is that by making poor food choices, you are in danger of permanently altering the bloodline, condemning your issue to an eternity of ill health. Clearly they don't feel there is enough guilt in the world.

Here are a couple of other examples. The first is from 'alternative' 'medical' 'expert' Joe Mercola's awful fucking website:

Epigenetics is probably the most important biological discovery since DNA. And it is turning the biological sciences upside down . . . In fact, you ARE changing your genetics daily and perhaps even hourly from *the foods you eat*, the air you breathe, and even by the thoughts you think.

Or this from the quite extraordinary Natural News, which seems to claim that epigenetics can explain how a positive mental state heals us from disease. Note how they somehow manage to weave in a fear of unspecified toxins along the way:

Today, a new scientific understanding is changing the conversation of biology and providing unexpected answers. It states that the mind and how we perceive the world around us directly influences our biology. Our positive energy (happiness and optimism) heals our conditions while our negative (anger and fear) toxifies leading to numerous diseases. Epigenetic control encourages people to become proactive in their own health instead of self-programming ourselves to be fearful victims.

So, given some of its fans, I have to be careful in this section, lest I am accused of misattributing something randomly to epigenetics. In reality, the rise of epigenetics in the alternative health community as a way of explaining magical beliefs is all the more curious, because epigenetics is nothing new. It simply refers to any changes occurring in genes that do not require an alteration in the DNA sequence.

Epigenetics is fascinating. It explains much of how genes work, turning them from a list of instructions into something that can create complex multicellular life. Indeed, epigenetics is common to all complex organisms. If it did not exist, it would render all life unrecognisable.

As most people know, every one of our cells that has a nucleus holds all of the DNA required to create a complete version of us. Because of this, in each individual cell, only a certain number of these genes are turned on or 'expressed' at any one time. A liver cell will only use the genes needed to make a liver cell. A fat cell will only express the genes required to make a fat cell. It is all controlled by the process known as epigenetics. The way that this happens is complex and a rapidly evolving research area, but many of the processes have been known about for some time.

In one of the most important processes called methylation, small chemical markers become attached to certain parts of the DNA sequence. Depending on the situation, methylation can either silence or activate certain parts of the genome, leaving particular genes more likely to be expressed. So for any cell in the liver, the amount of methylation will encourage the expression of genes needed to make a liver cell. DNA strands are also wound into complex structures called nucleosomes, held together by special proteins called histones. The tightness of this winding is another factor that helps to control which particular genes are expressed. Even when I was studying biochemistry twenty-five years ago, we knew about and studied a number of these processes, but new ones are being discovered all the time, and epigenetics is indeed a hot area of research. To call it a 'new science', however, is stretching that point.

The expression of different genes within a cell can be influenced by environmental factors. So if you eat ice cream, the gene that codes for the right digestive enzyme will be turned on by an epigenetic process. This gene will be opened up somehow in response to a signal, do its protein-making thing, and then turn off once enough of the enzyme has been produced. Epigenetics really is that simple: switching genes on and off. Without epigenetic processes, all our cells would produce all the proteins coded for in their DNA, all of the time. If that was the case, we would all just be a gelatinous puddle of organic matter quivering on the floor. Epigenetics is the way in which our strands of DNA manage to become a functional, structured human being, and is the reason why we can respond so well to the world around us.

Where epigenetics has become even more interesting in recent years is the discovery that some of these changes might be heritable (passed down from parent to child). Indeed, many definitions of the word 'epigenetics' only refer to these heritable changes, perhaps causing some of the confusion. It used to be thought that all DNA methylation was wiped away when a sperm fertilised an egg, but in 2013, a study showed that some methylation can get through intact, coming from either the male or the female side.[7] This indicated that changes to the way genes were expressed could perhaps be passed down through the generations. The implication was that a parent's environment could affect the way a child's genes were expressed before that child was even conceived.

The idea that a behaviour or environmental stimulus might cause changes that could be passed down to the next generation was certainly exciting news for Creationists and other zealots. Many misunderstood it to such an extent that they declared Darwin's theory of evolution dead. Maybe giraffes had long necks because their parents stretched them reaching for high leaves, because 'epigenetics'. Blacksmiths doing hard physical work might pass the benefits of their muscular physique onto their sons, because 'epigenetics'. Obese people who eat too much ice cream might have obese children, because 'epigenetics'.

It has since been appropriated by numerous quacks as an explanation for much nonsense, largely because it appeals to that central tenet of all quackery, that we are in charge of our destiny. And even more powerfully, maybe we are also in complete control of our children's destiny. You can see why epigenetics has great appeal to moralisers and guilt traders. It is not bad enough that you made yourself fat, you are actually destroying the life of a child you haven't even had yet. And all because you couldn't resist that ice cream. Somebody think of the hypothetical children.

In many ways, this makes epigenetics the darling of the neoliberal movement, a vehicle for stigmatising anyone perceived to be making poor choices. This has not been helped by a number of scientists working in the area, keen to make their field seem important and earth shattering. Here is an example from a perfectly respectable American biologist, Dr Randy Jirtle: 'Epigenetics is proving we have some responsibility for the integrity of our genome. Before, genes predetermined outcomes. Now everything we do – everything we eat or smoke – can affect our gene expression and that of future generations. Epigenetics introduces the concept of free will into our idea of genetics.'[8]

If this sounds far-fetched, that's probably because it is. Epigenetics is important, but when it comes to obesity, it is not quite the game changer some make it out to be. And the integrity of our genome is certainly not at risk from our diet.

THE *HONGERWINTER*

There is some evidence that epigenetic effects might have a limited influence on body weight. Towards the end of the Second World War, a German army blockade led to a devastating famine in the western part of the Netherlands. In the desperately harsh winter of 1944–45, already dwindling food supplies reached crisis levels as the German army destroyed ports and bridges in an attempt to halt the advance of the Allied forces into occupied Holland. The result was a brutal famine affecting millions, something that was to become known as the 'Hunger Winter' (*'Hongerwinter'* in Dutch). By February 1945, adult rations had dropped below 600 calories per person. By the end of the siege in May 1945, over 20,000 people had starved to death, with millions more undergoing terrible suffering.

The unusual thing about the Dutch famine is that it occurred entirely within a developed nation, with high levels of literacy and a good standard of record keeping. It also happened in a country that returned to relative normality shortly after the famine had ended. This meant that there was an opportunity to perform detailed epidemiology on the long-term effects. In 1976, a study showed that the sons of women who had become pregnant towards the end of the Hunger Winter were more likely to become obese than men whose mothers conceived shortly after it.[9] Studies in animals had shown that the offspring of starved mothers often had larger appetites caused by changes in the systems that regulated their food intake, and it was suggested that the same might be occurring in these *Hongerwinter* children.

It appeared to show that when mothers starved, their children had a tendency to become hungrier and fatter. Could it be that this was due to an epigenetic effect, caused by food shortage and passed down through the generations?

Supporting this hypothesis, children conceived during the Hunger Winter were shown to have differences in the methylation of certain genes, notably one involved in the creation of IGF2,[10] a growth-regulating hormone. IGF2 is thought to have a role in foetal development, and although it is not known whether or not it influences the body weight of adults, the changes in methylation are an indication that the epigenetic effects of starvation were being passed down from mother to child. It is not hard to imagine how this sort of adaptation might be useful. A child born into a period of shortage might be given an epigenetic head start, with adaptations to ensure that it stores as much energy as possible.

Although this particular epigenetic effect struggles to explain population-level increases in obesity throughout the twentieth century, some have suggested that our societal fear of fatness might have contributed to increasing numbers of pregnant women engaging in dieting and restricted eating behaviours. There is a chance that this may have passed on epigenetic adaptations to their children, perhaps inclining them to future weight gain. This is of course highly speculative, but underlines the importance of good nutrition during pregnancy, and how vital it is that women are not made to feel guilty about their dietary choices at that time.

CHILDREN OF THE FAT RATS

Starvation appears to be one driver, but there is also some evidence of a reversed epigenetic effect, at least in rats, suggesting that overfeeding might also pass down adaptations through the generations. If rats are fed exclusively on a high fat diet, their offspring become heavier even when fed normally, suggesting some sort of adaptation caused by their parents' eating habits before conception.[11] Although interesting, the findings have yet to be investigated fully in humans, perhaps because of the obvious ethical constraints involved in doing that work. Giles Yeo explains: 'If you isolate the sperm or eggs of mice on a high-fat diet and use IVF to produce offspring, they tend to be heavier. The effect is highest in the mother, but also happens in dads. But when it comes to human studies, the problem with epigenetics and obesity is that changes are unique to the tissue they are expressed in. The drivers of obesity are in the brain, and we cannot easily look at brain tissue in humans.'

It is also worth noting that although epigenetics appears to be some sort of mystical effect that science cannot explain, it is anything but. The mechanisms that cause epigenetic changes are all coded for in our genes. They have doubtlessly been selected for at some point in our evolutionary history, perhaps conferring an advantage within a particular environment.

Although epigenetics is an exciting area of study, the effects of the Hunger Winter are small, and the result of an extreme event. The changes seen in rats have not yet been observed in humans, and seem to disappear after one or two generations.

Certainly the way our genes are expressed is highly dependent upon our environment, and in some cases these changes can be passed down to our children. Some methylation is preserved when a sperm fertilises an egg, but it is important to remember that most of it is not. Heritable epigenetic effects are likely to be of little consequence when compared to the huge influence of our genes, the main mechanism by which characteristics are passed down through the generations.

ASSORTATIVE MATING

If our genes have hardly changed, and epigenetic effects are only small, presumably that means that our environment must be responsible for almost all the recent changes in BMI. This is a widely accepted truth in obesity science, and although most non-geneticists will grudgingly agree that interesting insights can be gained from looking at our genes, they do not explain why people are getting fatter.

This is the received wisdom, but there is a pretty big assumption being made by anyone saying it. Assumed truths are often the enemy of discovery, so it is perhaps worth exploring these assumptions in a little more detail.

Remember the ob/ob mouse? It was created by scientists selectively breeding mice over a number of generations in order to produce strains that become fat in a normal environment. The easiest way to do this is to take the fattest males and females and breed them together, then select the fattest mice from that litter and breed from them. You will probably want to encourage the fat mice to produce as many litters as possible, in order to give you plenty of fat babies to choose from. In a few generations, fatness is likely to become far more prevalent. Eventually, odds are that you will have produced a few extremely fat mice.

This sort of selective breeding can produce swift and dramatic changes in body shape and size, as well as profound behavioural adaptations, depending upon what is being selected for. Great Danes and pugs are the same species, yet selective breeding has made them profoundly different from their forebears. Chickens, cows, pigs, sheep and turkeys are hugely altered from their wild ancestors, and continue to change through selective breeding programmes. Although evolution by natural selection is often a grindingly slow process, intentional selective breeding is a high-powered rocket of genetic change.

As far as I am aware, no one has been selectively breeding fat people, although this probably has legs as a YouTube conspiracy theory. But when anyone assumes that genetic changes have not had a role in the obesity epidemic, they are also assuming that BMI has no influence over our choice of partner, and on the number of children couples produce. As it turns out, both these assumptions are wide of the mark. People with high BMIs are far more inclined to have children with similar partners, and larger couples have been shown to be more likely to have a greater number of children.[12] [13] [14] [15] [16] With these two pressures, it is possible that there is an amount of 'selective' breeding occurring – unintentionally on the scale of governments and demography, but totally intentionally on nights out and on Tinder. Larger people go for larger people, and tend to produce more children when they do. With obesity being predominantly determined by genetics, there is a distinct possibility that the various genes related to higher body weights will become grouped together in a small number of individuals, perhaps helping to create the growing tail on the right of the bell curve. The likelihood that people will be attracted to partners of a similar shape and size might just be helping to create increasing numbers of extremely obese individuals.

Assortative mating, the idea that we are attracted to people of a similar body shape, is a well-known driver of partner choice. But as we increasingly find our long-term partners later in life, and the average age of parenthood increases, it is perhaps likely that we are choosing mates at an age when we are more fully developed into our final body size. When marriage at sixteen or seventeen was common, the genes inclining people to a larger size were perhaps not fully expressed, making gene combinations more random. Could it be that later partner choice has created a selective breeding environment? And has this been compounded by our increasing stigmatisation of fat people, with larger people being excluded and marginalised from the rest of society, and so more likely to have children together?

Either way, the idea that assortative mating might be driving some of the worst effects of obesity does raise some interesting questions. If recent changes are just an inevitable consequence of people choosing a partner later in life, what does that mean for society? We surely don't want to return to higher levels of teenage pregnancy. Perhaps it means accepting that there are genetic differences that make some people look a certain way, and get on with trying to improve their health.

Framed in this way, obesity is a different problem, which is why these effects are rarely mentioned in the debate. The implication that it might be something we cannot change is perhaps too frightening for us to consider.

WHAT ABOUT THE 30 PER CENT?

Genetics plays a significant part in the prevalence of obesity, accounting for around 70 per cent of the reasons why someone might become obese. And with the influence of assortative mating, it even has the potential to influence the tail of the bell curve. The shifting of weight distribution in the population shows that different people are reacting to the changing environment in different ways, indicating that genetic variation is almost certainly a major factor. Katherine Flegal, the epidemiologist who first identified changes in US BMI distribution, said this in a 2012 interview:

> The distribution is becoming much more skewed. The lower end of the BMI distribution in the population is pretty similar over time, but the higher end is considerably higher. This suggests some kind of effect that varies by BMI level, possibly a gene-environment interaction, where those who are not genetically susceptible to the effects of the environment are little affected. However, it could also suggest social or personal factors that interact with the environment as well. Some groups may be socially more protected from the effects of the environment, with its ample, cheap food supply and the limited need for physical activity, than other groups.[17]

There are of course many factors that affect body weight, and we shall be investigating them in the coming chapters, but the role of genetics rarely gets much attention. Our understanding of genes doesn't really inform our public health strategies, because the prevailing belief is that they cannot account for rapid population-level change. Giles Yeo often jokingly describes himself as a 'bad person' when speaking publicly to health professionals, because in suggesting a role for genetics in obesity, he is perceived as absolving people of personal responsibility. Jeffrey Friedman has also been similarly criticised, and received much vitriol and derision after publishing an article in *Newsweek* stating that obesity is 'not a personal choice'.[18]

The science that underlies the influence of genetics on obesity is well established, but this rarely seems to filter down to the level of public policy. Take this statement from the New York State Department of Health, in an official report on obesity:

> The rapid changes in obesity prevalence over the past thirty years cannot be due to genetic changes, which take thousands of years to manifest.

Or this similar statement from a report compiled by the US National Bureau of Economic Research:

> Genetic factors cannot account for the rapid increase in obesity since 1980 – these factors change slowly over long periods of time.

Or this, from a National Institutes of Health Program Announcement:

> While there is certainly an important genetic component to obesity, the recent epidemic of obesity cannot be due to genetic changes in the population and therefore must be due to changes in environmental influences.

Similar views and statements can be found everywhere, both in academia, the media and government, often followed by reports that completely dismiss twenty years of groundbreaking research into the field, and a wealth of information that could genuinely help people.

It is a great shame that more attention is not paid to the insights from genetics when trying to design interventions to improve people's lives. Public health approaches are designed in much the same way as seatbelt laws, prohibiting certain behaviours through legislation, then encouraging compliance with advertising and behavioural nudges. But if we accept that obesity is largely a consequence of instinctive primal behaviour, and that the way we react to food environments is hugely individualised, then we must be careful. Although everyone is likely to be damaged by their face hitting a windscreen at high velocity, the health effects of different food environments are far more complex and nuanced. Just like the pasty Europeans who fried in the sun when they moved to Australia, there is a chance that some of us will react badly to these brave new worlds.

9

IS IT BECAUSE OF OUR GUTS?

If you want to know why people are getting fat, it makes sense to spend a little time looking at our guts. Although behaviour controls what we decide to put into our mouths, how much nutrition enters into our bloodstream depends largely upon what is happening in the long system of tubes running from our mouth to our butthole. Food can only be metabolised and stored as fat if it is absorbed, and this is largely controlled by our gut.

Obviously, if we want to use our guts as a way of explaining recent rises in obesity, we run into many of the same limitations we saw when discussing genetics. The physical structure of our guts is unlikely to have altered significantly in recent years. The way in which we absorb food was almost certainly determined a long time ago in our evolutionary history, with little chance of selective pressures changing it much over the last 100 years.

But that is not the whole story. Because rather than being static and fixed, the human gut comprises some of the most dynamic and changeable organs in the body. The reason is that the gut, and particularly the colon, is home to a rich, diverse community of micro-organisms, living inside us and feeding upon what we have eaten. Crucially, these micro-organisms, collectively known as the gut microbiome, can respond very quickly to environmental change. Their numbers and diversity vary enormously depending upon our diet, our health and our exposure to drugs such as antibiotics. Increasingly, evidence of how these tiny organisms might affect us in profound and surprising ways is emerging.

As with epigenetics, the microbiome is an exciting area of progress and discovery, making it rich ground for exploitative quacks. Even within academia, the importance of the microbiome is often oversold by excessively keen researchers. Jonathan Eisen is a professor of evolutionary biology who has spent the last ten years studying and researching the microbiome at the University of California, Davis. On his blog The Tree of Life, he gives out awards to other academics and media organisations he deems guilty of 'Overselling the Microbiome'. Most are awarded to people mistaking association with causation, so often at the root of misunderstanding.

For instance, recent studies showing that people with different emotional states have an altered microbiome were widely reported as evidence that gut bacteria influence human emotions. This prompted Professor Eisen to write: 'No no no no no and no. They do not show ANY connection between our thoughts and our microbiome. They just report a correlation. It could be that people with different thought patterns eat differently. Or people with different thought patterns exercise differently. Or $(#($(#@@ just about anything.'

To make matters worse, in recent years, the microbiome has started to become big business. Diet books that promise to leverage new understandings of the importance of our gut bacteria have flourished, as have testing companies that analyse stool samples and offer tailored dietary advice. Every condition under the sun has been attributed to alterations in the microbiome, with strong and sometimes surprising evidence of correlation in many cases. Although there is truth among it, as usual the problem is sorting out the wheat from the chaff.

As interest in the microbiome grows, and the gold rush continues, Professor Eisen is likely to be a busy boy – especially when it comes to the study of obesity.

WHAT ACTUALLY IS THE MICROBIOME?

Like almost all complex life, we share our bodies with a remarkable number of micro-organisms. These comprise an astonishing diversity of bacteria, viruses and yeasts, all living upon and within us. Take a quick look at one of your hands. At any one time, there will be more bacteria living on its surface than there are people on the planet.

At odds with the prevailing view of germs as dangerous agents of contamination and sickness, for the most part we live in a happy synergy with these invisible hitchhikers. Under normal circumstances they do us no harm at all, and there is increasing evidence that many of them do us a great deal of good.

It is often said, both in the popular press and in the abstracts of numerous scientific papers, that the average human contains ten times more bacterial cells than human ones. If you think this figure seems astonishing and hard to believe, then you might just be on to something, because it is almost certainly false. It is based on a rough estimate made in 1972 by the microbiologist Thomas Luckey, something that he probably never envisaged people would still be repeating over forty-five years later. Luckey based his estimate on an assumption that the distribution of bacteria was constant throughout our guts, and similar to the amounts per gram he had observed in stool samples. In reality, he overestimated the density in stools, and was not aware that bacteria are far more concentrated in the colon, with far lower levels in the intestines, stomach and rest of the digestive tract. But as we have already learned, people seem to like higher estimates of anything, and the ten to one figure stuck for many years.

Although a little less impressive, the reality is still quite remarkable. It is surprisingly hard to measure, but the most recent estimates place the ratio at around 13:10,[1] meaning that we have slightly fewer human cells than bacterial ones. Most adults contain around 30 trillion human cells and 39 trillion bacterial ones, although the bacteria are far smaller and lighter. We are still mostly human when it comes to weight, and just less than half of our cells are derived from human DNA.

The colon is the final section of our gut, a five-foot piece of tubing that connects our small intestine to our anus. It is the section where most water is absorbed and where the remaining bits of undigested food are converted into faeces. It contains our richest and most diverse colonies of bacteria, and is very much the home of the gut microbiome. In total, our gut microbiome is thought to weigh around 2 kilograms, with the large majority of that being located in the colon. There are over a thousand different species capable of colonising it, and the average human will have well over 160 types of micro-organism present at any one time. This gives the gut microbiome an incredible 150 times more unique genes than the human genome.[2] Fingers crossed, this seriously impressive figure is unlikely to be debunked as an overestimate any time soon, and gives an indication as to how potentially important the microbiome is when it comes to making humans more adaptable than our genes alone might allow.

It has become increasingly clear in recent years that the bacteria colonising our gut are not just unwelcome parasites, hitching a free ride and feeding off scraps in the lower regions of our gut. It now seems that the microbiome is hugely important for our well-being, helping us to digest food, and exerting an influence on a number of health outcomes. The vast majority of bacteria living within us are helping to process food that would otherwise be nutritionally useless, and to fight off pathogenic organisms that might do us harm. In return we give them a warm place to live, and a plentiful supply of nourishment.

When the genes contained within the gut microbiome are analysed,*[3] it appears likely that it is performing a number of functions that go beyond the simple conversion of food into poo. Complex sugars such as fibre, cellulose and pectin, all components of plant foods, are vital for its function, and much of the work of the microbiome appears to be in converting these into short chain fatty acids, which are then absorbed through the gut wall and into the bloodstream. Short chain fatty acids can then be used as an energy source, but they are also thought to have many other important roles, particularly in the functioning of heart, muscle and brain cells. Importantly, they seem to contribute to the reduction of inflammation, suggesting they may protect against a number of serious diseases.

The microbiome can also make certain amino acids that we cannot produce within our own cells, and produces a number of important vitamins. It is worth noting that the production of many of these molecules does not occur within a single type of bacteria. Instead, different bacteria undertake different parts of the work, the whole microbiome acting together in a sort of microscopic production line. A diversity of species is therefore extremely important for the microbiome to function successfully.

* Most of our gut bacteria can be split into two main types. The most prevalent is Firmicutes, with around 250 different species, and most of the others are Bacteroidetes, a group of around twenty different species.

Everyone's microbiome is different, and they are thought to vary significantly around the world. This is partly because people pick up different species from their environment, but it is perhaps more strongly influenced by dietary patterns. African children, for instance, have far more of the Bacteroidetes group than European children, and their microbiome contains a much greater number of genes for digesting complex sugars. This is presumably because they have a diet that is richer in fibre and complex plant foods, and their microbiome has adjusted in order to get the maximum benefit.

The microbiome is certainly not fixed. Dietary modifications produce rapid changes in the diversity and types of bacteria contained. Changing from a largely meat-based diet to one comprising mainly plant foods has been shown to alter microbial diversity within a single day. Animal studies have shown that changes to the diet account for most of the alterations seen in the microbiome. In humans, it seems that high-fat diets favour the Firmicutes type of bacteria, and diets with plenty of vegetables lead to higher levels of the Bacteroidetes type.[4]

CAN YOUR BACTERIA MAKE YOU FAT?

There is certainly some evidence that obesity is associated with low levels of bacterial richness.[5] In the world of tabloid headline writing, that would be enough to state that having too few bacteria makes you fat, but clearly, making any statement to that effect would put me in danger of winning an 'Overselling the Microbiome' award. Association alone should never be taken as proof that one thing is causing the other. For instance, it could be that being fat leads to people having less diversity of bacteria in their guts. Or perhaps a diet that makes people fat is also bad for bacterial diversity. Or maybe getting lots of exercise increases microbial diversity, along with decreasing the risk of obesity. Or maybe smoking causes high levels of bacterial diversity and promotes weight loss. There are so many potential confounders that it is almost impossible to make a straightforward link.

Perhaps for this reason, for many years the intriguing correlation between obesity and bacterial diversity remained just that, a correlation. But in 2004, the microbiome researcher Fredrik Bäckhed decided to try what we can only assume was the world's first poo transplant. He transferred the microbiome of an obese mouse into the gut of a mouse that had been raised in a sterile environment, without any gut bacteria at all. He found that the previously thin recipient quickly became obese, something that did not happen when it received the microbiome of a normal weight mouse. The experiment seemed to show that a specific mix of bacteria was making the mouse fat, suggesting a causal role for the microbiome in obesity.

Then, in 2014, a study of identical human twins of different weights found large variances between their microbiomes.[6] Although this alone was only proof of an association, when microbiome samples from these twins were transferred into the guts of previously germ-free mice, the mice receiving the heavier twin's bacteria gained significantly more weight. Could it be that something about the heavier twin's bacterial mix was actually causing weight gain? Was this proof that the microbiome causes obesity?

There have also been some tantalising results in human faecal transplants, where patients are given new bacteria from the faeces of a healthy individual to colonise their guts. This fairly unpleasant sounding process has been shown to be an effective treatment for persistent Clostridium Difficile infections, and is currently being tested for a number of other gut-health conditions. The suggestion is that seeding the gut with a healthy set of bacteria could reset the microbiome, and put the recipient on a path to better health. Treatment requires the recipient to swallow capsules of someone else's poo, usually freeze dried, and hopefully without chewing, or have donor poo inserted up the other end. This has shown promising results in a number of often debilitating conditions, things that are clearly unpleasant enough to make eating someone else's shit an option worth considering.

In a 2015 case study, a previously normal weight patient became obese after receiving a faecal transplant from her overweight daughter.[7] This suggested a role for her newly altered microbiome in weight gain, and echoed the effects previously seen in mice. Of course this was more anecdote than evidence, but around the same time a report from a clinic that had been treating patients using faecal transplants from normal weight donors found some evidence of weight loss in the year following the treatment. These results provoked a great deal of media interest, no doubt helped by the poo-eating angle, and added to speculation that the microbiome might have a significant role in obesity, and potentially as a weight loss treatment. But it is worth noting that when controlled trials have been conducted using faecal transplants from lean donors, although there does appear to be an influence on insulin sensitivity, no effect on weight loss has been seen when compared with placebo treatments (presumably some sort of 'fake poos').[8]

Although the evidence is conflicted, gut bacteria does appear to have a limited effect on weight gain, perhaps playing a role in the development of obesity. Providing this is true, exactly how might it be happening?

Kevin Whelan is a professor of dietetics at King's College London who specialises in the microbiome. He is more circumspect than many about the role of our gut bacteria in obesity, perhaps because he fears the ignominy of an 'Overselling the Microbiome' award. But mostly, I suspect, his reticence is because he does not have a diet book or testing kit to sell. He told me:

> The problem is, diet is often missed out of the discussion, and it is obviously really important. The relationship between diet, the microbiome and obesity is a mutually dependent one. Both your diet and your microbiome can affect your body weight and can of course affect each other also – it doesn't just go one way! This makes it really difficult to establish a direct causal relationship. There are a few trials of probiotic supplements, and a meta-analysis of these showed that people on a weight loss diet lost more weight if they also took a probiotic.* But the difference was only about 0.59 kilograms. The problem is that 0.59 kilograms might be statistically significant, but it is not really clinically significant for someone who is obese. There does appear to be a strain-specific effect, so the type of bacteria you are taking might be important, but it might well have more to do with diversity than any specific types. But saying that, even the importance of diversity has been challenged recently.

In fact, studies on obesity and the human microbiome have shown wildly inconsistent findings. Some studies have shown obesity to be associated with a high level of Firmicutes bacteria, some with high levels of Bacteroides. Others have shown no association at all. Obesity

---

* Probiotics are micro-organisms that may be of benefit when consumed. A prebiotic describes any food that might feed our gut bacteria, particularly fibre.

has been shown to be associated with a high diversity of bacterial types, but also with a low diversity. Despite studies on microbiome transfer in mice first showing an effect as far back as 2004, we are still no closer to knowing what an obesity-causing microbiome might look like, or what specific strains are involved. The complex interactions between diet, obesity, our bodies and our gut bacteria make it nearly impossible to assess exactly what is controlling what.

It seems likely that, just as there is very wide genetic diversity regarding how we respond to food, there is perhaps equal diversity as to how we respond to different microbiomes. The strains involved might affect different people in hugely different ways. And if diet can change the microbiome dramatically in just twenty-four hours, is the introduction of a particular bacterial strain ever likely to have a profound and lasting effect?

Kevin Whelan thinks that 'any effects are probably not due to one bacteria. It is more likely to be a number of bacteria working together. People spend a lot of time looking for a single strain that has an effect, and it is unlikely they will ever find one. It is similar to obesity genes. There are many that have a small effect on their own, but interactions between them can produce something significant.'

Although introducing a new microbiome to a germ-free mouse is a clever experiment with intriguing results, it is hardly representative of what is going on in the real world. It allows a level of control that does not normally exist, leading us to observe a one-way interaction, rather than the complex feedback mechanisms that usually occur. Although there is good evidence that the microbiome may contribute to an excess absorption of calories, or to inflammation and metabolic disorder, little is known about how and why this occurs. The microbiome, like obesity, is complex – it's not one simple thing that is going to magically solve our problems. But as Kevin Whelan says, 'To be honest, I am just a scientist trying to get the best, evidence-based information out there. Sometimes that is not what everyone wants to hear.'

CAN WE BLAME THE MICROBIOME?

For the purposes of this book, the key question is, have recent changes in the microbiome led to rises in obesity? The short answer of course is that we do not know. It is almost impossible to assess if gut bacteria have changed over time, particularly because no microbiomes from the past have been preserved. Even if we wanted to look at changes over the past twenty years, we are likely to run into problems. Part of the problem, Kevin Whelan says, is that our methods have changed. 'When I started my PhD in 2000, we used Fluorescent In Situ Hybridization, known as FISH. Now, no paper uses FISH, we use whole genome sequencing and metagenomics. There are no studies to compare the microbiome from twenty, thirty, forty years ago to how it is now. The best you can do is to look at old literature, but the methods used are archaic now, and you can't really compare.'

Still, this does not stop people from trying. Speculation that our modern diets, particularly consumption of highly processed industrial foods, has led to obesity causing changes in our microbiome are rife. Tim Spector is a professor of genetics and author of *The Diet Myth*, a book that explores the importance of the microbiome on weight loss, and he has had plenty to say on this subject, mostly in publicity interviews to accompany his book launch.

In order to explore the effect of modern processed foods on health, he conducted a high profile 'experiment' on his own teenage son, feeding him nothing but McDonald's burgers and nuggets for ten days, then observing changes in his gut bacteria. He found that there was a 40 per cent reduction in the diversity of the species present, leading him to conclude that the processed industrial foods his son was consuming had caused the changes. Helpfully for someone promoting a book, this was reported in many newspapers and magazines, with Professor Spector speculating that 'this could well have been due to the surge of chemical additives, as well as the effects of nutritional and fibre deprivation'.

Clearly, feeding his son McDonald's for ten days was a shameless publicity stunt, and little should be read into any microbiome changes. It is not surprising that placing someone on any heavily restricted diet with only low levels of plant-based food might cause a drop in bacterial diversity. Almost no one eats one meal over and over again without any variety whatsoever. On top of that, to do this without any control group is a misrepresentation of what constitutes scientific enquiry, and it is a shame that it received so much coverage.

Professor Spector has suggested a role for chemical additives in the development of obesity and many other metabolic diseases, largely due to their supposed effect on the microbiome. This view is supported by many other commenters, although few with Professor Spector's academic credentials. Here are some typical extracts from newspapers, books and websites, perhaps deserving of an overselling award.

> Eating junk food kills stomach bacteria which protect against obesity, diabetes, cancer, heart disease, inflammatory bowel conditions and autism.
>
> Luke Heighton, *Telegraph*

> Whether you are fit or unwell, overweight or slim, happy or depressed, I believe we can all benefit from taking the best care of the armies of microbes in our gut. If you follow my plan – which will be featured all next week in the Mail – you should experience fewer cravings, lose excess weight, and hopefully reduce bloating, wind and gut pain.
>
> Michael Mosley, author of *The Clever Guts Diet, Daily Mail*

> Here's a plan that will show you how to starve the fat-forming bacteria, reseed your gut with good fat-burning ones, and fertilize those friendly flora with just the right foods to reboot, rebalance, and renew your health – and lose weight for good. It's all based on up-to-the-minute scientific research.
>
> Gerard Mullen MD, author of *The Gut Balance Revolution*

It is true that there is some evidence that emulsifiers, common additives used in many different foods,[9] have an effect on the gut microbiome and metabolic health of mice, leading researchers to conclude that this might be contributing to 'increased societal incidence of obesity'. But it is worth noting that the study was in mice and involved very high doses of emulsifiers, unrepresentative of what people are likely to eat in daily life. Despite the numerous headlines this study generated, similar metabolic effects have never been observed in humans, on whom the same emulsifiers have been extensively tested.

In truth, many things can alter our microbiome, and there is a great deal of concern that an overuse of antibiotics might be having a detrimental effect. It is also frequently implied that our modern obsession with hygiene and cleanliness has led to microbiome disruption as constant use of soaps and hand gels means fewer chances for different bacterial species to colonise us.

Although both of these are valid concerns, and the overuse of antibiotics is troubling for other reasons, any role in the development of obesity is still extremely speculative, especially considering how little is known. Antibacterial hand gels and modern drug treatments have brought down the incidence of gastric problems, something that may have led to enforced periods of fasting in previous generations, and so perhaps reduced levels of obesity. But would we really want to go back to a time when gastric disorders were one of the most common causes of childhood death? And are we really sure that most people's microbiomes are less healthy than they were in the past? Or are we just projecting a fear of modernity onto our invisible gut friends?

WHAT DOES IT REALLY TELL US?

Given the many adaptations that we have already seen, it's an interesting idea that a less diverse microbiome might be a reaction to becoming fat. After all, our bodies have evolved numerous other mechanisms to resist weight gain, and many more to resist weight loss. But we should not be too surprised if we find that these mechanisms are not universal, since many aspects of the microbiome are influenced by our individual genes.[10]

The microbiome offers us a rapid adaptability that our genes cannot provide, something that has clearly contributed to us evolving in synergy with it over many years. As food availability changes, so does our microbiome, allowing us to adjust our response to a wide variety of dietary patterns. Evolutionary changes may have enabled early humans to adapt to new patterns of food availability within a few generations if the pressure was strong enough. But the dynamic nature of the microbiome allows such changes to occur within a single day. This would have given early humans the ability to cope with and colonise many different environments, and to deal with significant seasonal changes in food availability.

It is entirely possible that some obese people have fewer bacteria as an adaptation to being obese, leading them to extract less nutrition from food, and so helping to control excess weight gain. But it is also possible that in a different obese person, specific adaptations cause them to extract more nutrition from their diet, so leading them to further gains. It has been shown that when some people restrict their dietary intake, their bacterial diversity quickly increases,[11] which is perhaps just one of the many responses to starvation we have already seen. Every aspect of these interactions is complex and individualised. Despite what the microbiome diet books might tell you, there is no one size fits all solution.

One day, techniques may be developed to enable microbiome-based treatments that have an influence on our health, body weight and perhaps even behaviours. It does appear that there is an effect, and even if this is small, it is potentially important. But right now, there is no evidence that specific eating advice tailored to gut health alone is likely to have a role in weight management.

SO WHAT SHOULD WE DO?

Although I am troubled by Professor Spector's sweeping statements about chemical disruption from processed foods, and by his overex-trapolation of pointless, publicity grabbing McDonald's 'experiments', his advice on how to eat is largely sensible. He has railed against calorie restricted weight loss diets in the past, saying that 'it's not what you should cut out, it's what you should eat instead. This obses-sion about fructose or certain types of saturated fats is missing the point.'

All the evidence about the microbiome can currently tell us is that a varied, diverse diet is the best advice when it comes to good health. Happily, this is also the best advice when it comes to eating well. We should love what we eat. Fear-based messages lead to restriction, which is bad for us. A burger or a chicken nugget may not be the best thing we can eat, but in attributing a morality to that choice, or demon-ising it as somehow toxic, it gains a significance beyond any effect it might have on health. Certainly, if burgers are all you eat, that is a poor diet. But I can guarantee that if I restricted my diet to nothing but kale and cucumber, my microbiome would be similarly diminished in a short time. Surely the only sensible approach for eating a varied diet is to ditch any rules of restriction. We shouldn't idolise or revile any single choice. We should allow anything and encourage everything. On the smorgasbord of foods available to us, burgers and nuggets are just two.

A varied and interesting diet will only ever include those choices occasionally, and there is no evidence that this will harm us.

We currently know precious little about what an obesity-preventing gut microbiome looks like, or how to achieve one. It may well turn out that any specific advice is fundamentally flawed, because what is healthy for one person is completely the opposite for someone else. In the midst of all this uncertainty, it seems that increasing the diversity of our diet is the only sensible advice. That way we spread our bets, while enjoying plenty of different foods along the way. Fibre and complex carbohydrates seem to have a positive effect on our gut bacteria, and the short chain fatty acids produced seem to be of some benefit. But other bacteria thrive when different foods are consumed, and who knows what the role for them might be in improving our health.

Kevin Whelan suggests a less prescriptive approach than the many people selling books and diet plans, perhaps because of a better appreciation of this uncertainty:

> I suppose the most important question is – would a standard weight loss diet be physiologically and nutritionally different to a microbiome-based weight loss diet? Is there a way to target the microbiome specifically? Really, to get to that point, we probably need to be cleverer about what is going to work in whom, and we just don't know enough yet. In general, unless something new comes up, we won't make any major inroads. It is a complex system, and without changing loads of other stuff, we won't solve everything using the microbiome alone.

The microbiome is undoubtedly important and interesting. It is worth noting that for a number of conditions, gut bacteria have been shown to have specific and important roles. New therapies are being developed all the time, and many of these techniques will change people's lives for the better.

But when it comes to obesity, the relationship is so complex and interdependent that specific therapies or diets are unlikely to have much of an impact. The advice remains to eat loads of different stuff, plenty of fruits and vegetables, a good helping of fibre and lots of wholegrains. Even as knowledge of the microbiome has increased exponentially, this advice has not changed. Professor Whelan recently published a study suggesting that wheat, legumes, garlic and onions may be helpful for a healthy gut. He says that 'the definition of healthy eating is a balanced diet. It is not about kale and coconut oil but about eating a wide range of fruit and vegetables, whole grains and healthy fats. However, it can be difficult to ensure you are getting a sufficient amount of fibre, which is key for a healthy digestive system.' Far from being the idealistic opinion of a food-loving chef, when it comes to our health, this is still the best advice there is.

10

IS IT BECAUSE OF CALORIES?

Portion sizes have grown. Snacks have become bigger. Once occasional treats are now being consumed every day. In our modern, privileged lives, we are free to roam the gardens of plenty. Every year the temptations on offer are bigger and better than before. Surely the real reason we have become fat is because we are eating too much food, and just can't stop ourselves. But I think we are far enough into the book to say that things are probably a little bit more complex than that. After all, the mechanisms that compensate for increases in food intake are many and varied, and, crucially, many of them operate over very long time-scales. If we feast today, our bodies often find a way to compensate tomorrow. Or perhaps next week. Or maybe even next year.

To further complicate matters, not every calorie we consume is equal. What we do with it, whether we burn it during exercise, store it as fat or use it to raise our metabolic rate depends upon many things. It may alter according to our age, size, sex and life stage. If you take two people who weigh 80 kilograms, it is unlikely that they will react in the same way after eating an identical plate of food. For instance, if one has just finished crash dieting all the way down way from 180 kilograms, and the other has just gained 20 kilograms, then the calories they consume will likely be used very differently.

In Part I we discussed how tiny changes to my daily food consumption might have caused me to gain significant amounts of weight over a twenty-year period. This was only to illustrate the ridiculousness of any assumption that my weight is fully within my control, but it is worth expanding the point here. Clearly, I rarely eat exactly the same number of calories as I burn in any single day. If I do occasionally eat to excess, my body compensates with extraordinary precision. Considering that calorie labelling on most food packaging is only thought to be accurate to within a 10 per cent margin, this compensation is far more precise than any nutritionist or dietitian could muster, and does not depend on me monitoring my intake. In reality, if I had eaten a couple of extra olives each day for twenty years, the odds are that I would still be the same weight as I am now. And similarly, I would probably be unchanged if I had eaten one olive fewer. The only reason I am not noticeably fat or thin is that the systems I have in place to maintain my weight are working pretty well. Within certain limits, the actual number of calories I have consumed has surprisingly little relevance.

When I mentioned in Chapter 2 that an excess of around 7000 calories would be converted into a kilogram of body fat, that was a rough but commonly used estimate. But as a predictor of weight change, it utterly disregards the many known and well-studied compensatory mechanisms our bodies use to keep our body weight in homeostasis. Even though the figure is deeply flawed, it is often used clinically and in the development of anti-obesity policy. Weight-loss advice from the NHS in the UK and the National Institutes of Health in the US suggests that cutting 500 calories a day should result in half a kilogram (a pound) of weight loss per week. These estimates are not just rough and ready, they are completely misleading. For most people, cutting that many calories will not induce the predicted weight loss, because our bodies will fight it all the way. In almost all cases, the calorie reduction required to shift half a kilogram is far greater, and will vary depending upon many factors. Overestimates such as these must be extremely disillusioning for dieters, who are often told that they are getting their calculations wrong, eating in secret, or just not trying hard enough.

Similarly, when predictions about the effect of policy interventions are made, they often use this figure to demonstrate the dramatic difference the new scheme will make to the nation's waistlines. Numerous studies and reports on the potential impact of taxing sugar-sweetened drinks have used the 7000 calories per kilogram figure.[1] It is not surprising that these policies rarely have the predicted effects, often being abandoned as a result of this failure.

BIG NUMBERS SOUND COOL

The use of these flawed calculations is all the stranger because good
models do exist that predict exactly what effect calorie reduction will
have. For instance, Kevin Hall's team produced a summary of their
work on energy balance in 2011,[2] creating an online tool to produce
more accurate estimates.[3] Their models are based on current knowledge about how the body responds to changing intake, including many
of the hormonal responses covered earlier in this book. Unsurprisingly,
it predicts far less dramatic results, both at an individual and population level.

The models predict two very important things. Firstly, small changes
in calorie intake over a long period are not enough to explain large rises
in body weight. They suggest that in the US, the average increase in
body weight between 1970 and 2000, roughly half a kilogram per year,
can be accounted for by a daily increase in energy intake of around 250
calories per person. This is far more than the tiny changes predicted by
other models, usually around ten calories per day.

Unfortunately, it also means that in order to reduce rates of obesity,
far more dramatic interventions are needed. Calorie intake needs to
come down much further than previously thought. Using the 7000
calories per kilogram model, a US Department of Agriculture report
estimated that a 20 per cent tax on sugar-sweetened drinks could
reduce average body weights by just under 2 kilograms per year, leading to an almost complete reversal of obesity to pre-1980 levels within
five years. Kevin Hall's mathematical models suggest that people might
lose 2 kilograms in total over five years. They also predict that rather
than being a linear loss – with people losing the same amount of weight
every year – decreases would tail off, as people's metabolisms resisted
further weight loss as time progressed. Even if the number of calories
being consumed remained permanently low as a result of taxation,
eventually no further weight loss would occur.

But still we forge ahead because people like higher estimates. This
is especially true when it comes to justifying their favourite new policy,
even if that means disappointment down the line. For the patients
being judged for their lack of weight loss success, this can be very
harmful indeed.

THE CALORIE CONUNDRUM

If the 250 calories a day figure in the US between 1970 and 2000 is correct, then where exactly did all this extra food come from? Can it be accounted for by changes to the diet? Well, over that period, the availability of calories in the US food supply rose by around 750 calories per person per day, more than enough to cover this increase. At the same time, food waste rose by around 500 calories, leaving 250 calories over, exactly the predicted amount. This suggests that increases could be accounted for by changes to the food supply alone.[4] Case closed?

Known as the 'push hypothesis', the idea that people have gained weight because the food system drives them to consume is tempting in its simplicity. But it does rely on a few assumptions. Firstly, it completely discounts a role for changing levels of physical activity, something that we shall discuss in the next chapter. But secondly, and perhaps more importantly, it suggests that all the available extra calories are being consumed, eaten just because they are there. This would seem to be a self-evident truth, clearly seen in the growing waistlines around the world. But strangely, when we look a bit closer at people's diets, some surprising patterns emerge.

Although common wisdom tells us that we are eating more and more, and portion sizes are growing exponentially, some of the data tells a different story. The UK's dietary surveys are recognised as some of the best in the world, monitoring changes at both the family and the individual level.[5] [6] Although all self-reported food surveys are plagued with inaccuracies, with participants often wanting to appear more virtuous than they actually are, it is hard to doubt the significance and consistency of the evidence relating to how our energy intake has changed. All of the surveys seem to show that, while obesity has steadily been increasing, the number of calories being consumed in the UK has been gradually going down. In the 1950s, the average daily calorie consumption was around 2,600. The most recent estimate, using the same type of survey data, puts the current figure at around 1,800 calories.[7]

Over the same period, consumption of fat, saturated fat, sugar, carbohydrates and salt has also decreased. In fact, the only things that have risen significantly are protein, as we favour more lean meats and fish, and fruit, presumably due to increased availability out of season.

So unless we think that fruit and fish are the primary causes of obesity, some places other than our diets, which seem to have improved in almost every way, deserve to be looked at.

It's worth noting that figures from dietary surveys are laced with controversy, and different ways of measuring often produce different results. There is an argument that in the 1950s, an era before the desire for thinness was so pervasive, people may have overestimated how much they ate in order to pretend they were wealthier. And equally, recent survey participants might be inclined to replace the odd takeaway with a virtue-signalling salad, ironically lying to demonstrate their moral fortitude. But the findings of different diet surveys using many and varied ways of monitoring food intake are broadly consistent, and the downward calorie-consumption trend seems to have persisted through the decades. In the UK at least, as we have become bigger, we seem to be eating less and less.

Susan Jebb, population health professor and obesity expert, has spent a great deal of time analysing these figures, and has come to much the same conclusion, which she explained to me:

> The challenge in measuring dietary intake is one of the reasons why there is so much controversy. However, all the methods show some broadly similar things – total energy has come down, proportion of fat has decreased from about 40 to 35 per cent of energy, with slightly less saturated fat, protein has increased, and carbohydrate consumption is largely unchanged. There is a greater proportion of carbohydrate from sugar than twenty or thirty years ago, with maybe a very small decrease in that happening now.

This calorie conundrum has led many to claim that reduced rates of physical activity must be the cause of obesity, and Susan Jebb thinks there may be something in that idea.[8] Others believe that a quirk of the modern macronutrient breakdown is the cause, or perhaps some other mysterious factor hidden in our modern dietary patterns. But when it comes to calories, it does appear that something curious and unexpected is occurring. Because although it is true that too much food will make you fat, and equally the case that too few calories will make you thin, it turns out that this doesn't help us explain obesity.

THE PLATE SIZE DILEMMA

Even the most accurate and sophisticated mathematical models in the world will show that if we cut the number of calories people eat, eventually they will lose some weight. And so, policies intended to make people thinner tend to proceed by finding ways to cut calories from the diet. There is little doubting that this is a sensible strategy, although given the capacity for people to compensate for small changes in calorie intake, it has more limitations than some might admit.

There has been much recent focus on portion sizes, particularly how they have increased over the years, but also how decreasing them might encourage a reduction in total calories consumed. In 2015, a detailed meta-analysis of studies into portion size created a great deal of interest, suggesting that reducing the size of packaging, tableware and serving size could drop the UK's average calorie consumption by 200 calories per day.[9] This led to headlines such as:

Growing Portion Size a Major Factor in UK Obesity – the *Guardian*

Portion Size Key in Tackling Obesity – BBC News

Take Portion Sizes Back to the 1950s to Beat Obesity – *Daily Mail*

(Somehow even talking about food portions, the *Daily Mail* neatly found an excuse for a return to the past.) The review, endorsed by the Cochrane Collaboration,* was certainly a high-quality one, suggesting a beneficial effect of reducing portion sizes, and also of decreasing the size of the nation's crockery.

But even here, things are not entirely what they seem. Several of the studies covered in the review were from the laboratory of Professor Brian Wansink, head of the Food and Brand Lab at Cornell University, and a well-known figure in the world of food behavioural science. In late 2016, shortly after the Cochrane review was conducted, Wansink published a blog post entitled 'The Grad Student Who Never Said No'. In it, he outlined a study his lab had conducted in an all you can

---

* Now known simply as Cochrane, the organisation formed to help medical professionals by conducting systematic reviews of all the experiments conducted in a particular field and combining the results to give a more complete picture. Endorsement by Cochrane is seen as a mark that the research has been done properly and the results can be trusted.

eat Italian restaurant. One group of customers had paid half as much as the other to see if there was an influence on eating behaviour, but it had failed to produce the expected results. Instead of writing this off as an expensive failure, Wansink asked an unpaid graduate student to search through the data set in order to test different hypotheses and see if there was anything worth salvaging. Eventually, after much hard work, the student had three papers published as a result, all co-authored by Wansink. The blog post was intended to be a motivational one, highlighting the importance of grabbing opportunities with both hands when they arise. Other post-doc students had turned down work on the restaurant data set, claiming they were too busy. Yet the new student had reaped great rewards from hard graft and perseverance.

This might seem to be a good life lesson, and a sensible use of time and resources, but to many researchers in the field of behavioural science, it was tantamount to admitting misconduct. In the comments section, Paul Kirschner, a professor of educational psychology, wrote: 'Brian, is this a tongue in cheek satire of the academic process or are you serious? I hope it's the former.'

The problem most commenters found was that this sort of data analysis, where the data set generated by a failed hypothesis is analysed again and again from different angles, is highly likely to create false positive results. Once the original hypothesis of an experiment is found to be null, it is not considered good practice to return to it repeatedly, testing different ideas until something shows up. In any complex data set there is always the possibility of finding false correlations, and so, generally speaking, it is only when the hypothesis being tested is shared in advance of the study that firm conclusions can be drawn. Wansink's blog post seemed to be admitting to something known as 'p-hacking', looking for statistically significant correlations without previously defining what you are searching for. Perhaps even more troubling, a senior and highly influential researcher did not even seem to be aware that this practice was problematic. Wansink claimed never to have heard of the term 'p-hacking' until he was accused of it.[10]

Often dubbed the 'Sherlock Holmes of Food', Wansink was very much the superstar of his field, with publicity-grabbing research findings, involvement in the development of US government obesity policy, lucrative corporate consultancies, and popular books such as *Mindless Eating* and *Slim By Design*. His work is extremely influential, and much of it is written into food lore. Anyone with an interest in the reasons why we choose to eat or drink in a certain way will be familiar with the findings of his laboratory. When I was researching my first book, he was one of the first on my list of people to contact. Thankfully, he never returned my calls.

If someone says 'people make 250 decisions about food every day', that figure comes from Wansink. The popular snack packs of different foods, all limited to 100 calories, were developed based on the outcomes of his research. If you think that people are likely to consume more when they use short, wide drinking glasses, that's Wansink again. Men eat more in the company of women, the same yoghurt tastes better when it has an organic label, keeping cereal boxes on the breakfast table makes you eat more – all Wansink. If you ever move unhealthy snacks out of sight in your kitchen, and place a fruit bowl on prominent display to improve your diet, that too can probably be traced back to work carried out in Wansink's lab. And crucially, the commonly cited belief that people eat less when they use small plates comes largely from his work. Many popular 'Small Plate' movements and campaigns started on the back of this, encouraging families to downsize their dinner plates for their main meals. I am fairly sure that some of the people reading this will have invested in new crockery as a result.

Much of Wansink's research is based on the popular notion of 'nudge' theory, the suggestion that the food environment can be subtly redesigned in ways that will encourage people to make better choices. The idea of 'libertarian paternalism', where positive choices are influenced rather than enforced, has become a powerful political movement in recent years, especially in the development of food and health policy. Wansink was very much at the heart of this, and had influence at the highest level, both in the US and UK. He repeatedly spoke against taxation policies to influence patterns of food consumption, which, given his extensive corporate links, troubled many people. If there were significant problems with his work, this was serious news.

The blog post had sounded warning klaxons around Wansink's research methods, and cast doubt on some of the Food and Brand Lab's papers. But soon things were going to get much worse. Jordan Anaya, Nick Brown and Tim van der Zee, a group of researchers dedicated to uncovering statistical anomalies in scientific research, analysed the papers mentioned in the blog post using specially designed software and found a catalogue of potential errors.[11] Troublingly, as more and more of Wansink's papers were analysed, increasing evidence of inconsistencies started to emerge. Nick Brown also turned up evidence of 'self-plagiarism' in a number of articles,[12] where large sections of text appeared to have been copied and pasted from one publication to another. He also found some suspicious duplication of data across papers, which was potentially evidence of serious misconduct.

Wansink issued a statement regarding the allegations, admitting that errors had been made, and a number of the offending papers have since been retracted.[13] Although he introduced new standard operating procedures at the lab, an investigation by the university found he had 'committed academic misconduct in his research.' Wansink announced his resignation in September 2018, but doubts remain over the research findings of this highly influential figure, with many of the headline grabbing results proving impossible to reproduce when conducted in other laboratories, suggesting they are little more than anomalies. Yet by that time, the genie was out of the bottle.

The popular ideas of nudge theory are tempting to politicians, public health organisations and large corporations, not only because they are cheap, but because they provide an illusion of control. They suggest that health inequalities, deep structural problems and modern lifestyle diseases can all be countered with simple environmental changes. They provide the illusion that in rearranging a cafeteria, or redesigning a retail store, we can somehow iron out huge, societal-level problems. Perhaps this is why Wansink's research proved so popular, and his methods went unchecked for so long. I'd love to believe that we can all lose weight if we just put our cereal box in the cupboard, or cure diabetes if everyone buys smaller plates.

This brings us neatly back to plates and portion size. Four of the papers in the 2015 Cochrane review were authored by Wansink, casting doubt upon the findings. In response to queries about this issue, the authors of the study noted that, while Wansink was 'unable or unwilling' to clarify some of the data, they didn't think it affected the overall findings.

But despite this, research on how plate size influences how much we eat has proved extremely controversial. Eric Robinson from Liverpool University has been a regular critic, and in 2016 he conducted a systematic review that suggested plate size had no consistent effect on food intake.[14] He told me:

> The plate size thing is a bugbear of mine. It drives me crazy. The science behind it is fairly ropey . . . Behavioural nutrition is very complex, with a great deal of variability, so replication is always going to be difficult, and in many cases it isn't even tried. People end up wedded to their ideas, but the problem is, it can end up with the public saying 'is anything useful?' So when there are robust and reliable findings, like there are with portion size, sometimes people don't listen.

PORTIONS, PORTIONS, PORTIONS

Of course plate size is a side issue really, little more than a tempting proposition. All the people who did invest in smaller plates only wasted a little cash, and the crockery industry were probably grateful for the uplift (to be clear, despite Wansink's corporate links, I am not suggesting a conspiracy by the dark forces of 'Big Crockery'. Or even 'Big Small Crockery').

But the question remains, is cutting portion size a potentially useful strategy for weight loss, either at an individual or population level? It sounds like a winner, and there is some evidence that increases in portions have coincided with rises in body weight, but it is worth remembering that in the UK, calorie consumption actually seems to have fallen over the same period.

But as the earlier quote from Eric Robinson suggested, the research on portion size is fairly robust and replicable. Obviously, if people consume smaller portions they eat fewer calories, but this is only part of the story. The key question is, will they compensate for that reduction by eating more later on? And if a snack is reduced to two thirds of its previous size, will people just end up eating two of them?

Although there are clearly limits, it does appear that portion-size reduction might be an effective way of reducing calories. As Eric Robinson explains:

We have looked at how much you can reduce the size of a pasta ready meal before someone starts to compensate for the reduction in calories. We found that you could reduce it by up to 30 per cent, and people don't compensate later. We and others have looked over one meal and over one week, although no matter how far you go, someone will say, how about one month, or six months. But there is no evidence to suggest that these sort of small adjustments are compensated for.

Ciarán Forde is a sensory scientist and principal researcher at the Singapore Institute for Clinical Sciences. He studies some of the complexities underlying why people may or may not compensate for calorie reductions.

Over many years of experience with foods, starting in childhood, we learn to associate certain sensory experiences with indulgence and fullness. In this way we learn to associate sensations like 'thickness' and 'creaminess' with the delivery of energy, as oftentimes these sensations are associated with fat or sugar. Today, it is possible to replicate these sensations using low or no-calorie ingredients, and use this to support reductions in energy density, without a reduction in food enjoyment.[15] [16] [17]

He continues:

Calorie dilution is one approach, but we find the best results when we dilute calories, limit portion-size and slow eating rates, without reducing food palatability. In this way we can use food formulation and texture to control both energy density and eating speed, to control the rate of energy intake (kcal/g/min) or what we term 'calorie velocity'. It is possible to reduce eating rate using slightly harder or thicker textures and still maintain a food's hedonic appeal. Our studies show that a reduction of approximately 20 per cent in eating rate (i.e. 50g/min to 40g/min) results in a 10 to 15 per cent reduction in energy intake within a meal, with no significant reduction in post-meal fullness or re-bound hunger. For calorie controlled meals the main complaint is often feeling hungry afterwards, but by putting the eating experience at the centre of meal design, it is possible to encourage greater fullness per kcal consumed by optimizing the sensory and eating properties of calorie reduced meals.[18]

So even something as simple as how quickly we eat can make a significant difference to our enjoyment of food and how full we feel afterwards. And despite how it may seem, not all calories are equal.

IS A CALORIE JUST A CALORIE?

There is evidence that when our energy comes from protein, things might be slightly different. Not only does protein require slightly more calories to digest than other macronutrients, reducing the overall calorie load, it has the added benefit of being more efficient at reducing hunger.[19] This has led to a raft of new convenient high-protein products, making it the new nutrient of choice.

Generally speaking, high-protein convenience products taste awful, and are presumably eaten for the health halo surrounding them, rather than as a pleasurable taste experience. Ciarán Forde explains:

> Protein seems to be more satiating than carbohydrates or fat, but it is expensive and hard to make delicious. We see better results when we focus on pleasurable eating experiences based on food, rather than individual nutrients. You should not pick foods that you don't like. Anyone actively sacrificing enjoyment of food will end up with an aberrant eating pattern. Trying to exert constant control over your diet is likely to lead to failure, and a dysfunctional relationship with food.[20] [21] [22]

The view that different calories may be compensated for in different ways is certainly intriguing, but it does not stop with macronutrients, calorie density and portion size. Sometimes, the format in which energy is consumed appears to make a difference. Richard Mattes is a professor of nutrition science at Purdue University. In 2007, he published work that suggested energy consumed in a drink is not compensated for in the same way as energy consumed as food.[23] This has since formed the basis for much of the worldwide demonisation of sugar-sweetened drinks, and the popularity of taxation policies on them. Professor Mattes told me:

My view is that it is not the sugar in sweetened drinks specifically. It is just that we don't monitor energy from beverages as
precisely as energy from solid foods. Consequently, energy from
beverages is largely uncompensated. Throughout most of
human evolution, we would not have obtained much energy
from beverages. Just on the cognitive levels, most people feel an
apple makes you feel fuller than an equal amount of energy
from apple juice. Additionally, thirst is a more salient acute
signal than hunger and there is less consequence of overconsuming water than energy from a solid food. This enables the
drinking of beverages, which if they are high in energy might be
problematic. It is a common experience that if we want people
to put on weight, we typically provide them energy in a liquid
form.

Although a calorie is a calorie, the differences in how our body reacts
dependent upon the situation renders that statement a vast oversimplification. We compensate in different ways due to numerous cognitive
and behavioural factors, as well as a great number of physiological
effects.

And although the effects of portion size and format are undeni-
able, whether or not short-term disturbances in our calorie intake
are significant when it comes to long-term weight gain remains to
be seen. The processes that maintain and control the amount of fat
we have in storage are governed by far more than just our food
intake. They operate outside the timescales of many of the calorie-
compensation experiments discussed here, and have evolved to ride
out any short-term fluctuations in food availability with exquisite
precision.

As we have seen, restrictive dieting might work for a year, maybe
longer. But it rarely produces sustainable long-term alterations in body
weight. To really understand the effects, experiments over a five- or
six-year timescale would be required. These would be nearly impossible to run and might not help us much. Obsessions with calories,
control and denial are known to be damaging, perhaps even predictive
of long-term weight gain. Our strategies might have to be a little more
sophisticated than asking people to eat smaller meals.

Has an overconsumption of calories led to rises in body weight?
Could a control of portion size lead to a significant fall? In a limited

way, the answer to both these questions is probably yes. But there are so many caveats, it is hard to be too enthusiastic. In truth, if we spent less time worrying about energy in, and more time thinking about what is good for our health, we would be considerably better off. If we really want to address health inequalities around the world, we might need more than just smaller plates.

11

IS IT BECAUSE WE
ARE LAZY?

Eat less, move more. Despite all the insights from genetics and endo-crinology that utterly discredit it, this simple Hippocratic notion still dominates the obesity discourse. So, if we are eating less but still getting fat, presumably that means that we are not moving around enough. Remember, in the UK at least, the best available data shows that people are eating fewer calories than they were forty years ago and still gaining weight. It stands to reason that a lack of exercise must be the cause.

Certainly, many people believe that our increasing laziness is one of the main drivers of population-level weight gain. And many others think this belief exonerates the food industry and allows it to shift the blame for expanding waistlines. So Coca-Cola sponsors sporting events, claiming that an occasional sugar-sweetened drink is fine if you just burn off the calories elsewhere. And Cadbury gives away school sports equipment, suggesting that chocolate does no harm if kids are running and jumping their way to thinness.

Although it is often repeated that we are less physically active than we used to be, there is actually very little definitive evidence that this is the case. It is obviously very hard to judge, but anecdotally most people will tell you that fifty or sixty years ago, we were burning more calories in everyday life. These days, we drive everywhere instead of walking. Far more of our jobs are sedentary. Leisure time is increas-ingly spent staring at screens, with media companies creating increas-ingly clever ways to keep us all sitting down. Longer commutes to work make walking impossible, and working hours mean that we are often too exhausted to do anything other than watch TV or sleep when we get home. There is also a mass perception that the streets are not as safe as they used to be, meaning that many young people are forced to spend an increasing amount of their time indoors. Molly-coddled and hidden away, they interact with the world only through their phones and tablets. They sit alone in their bedrooms, greedily troughing snacks and growing fat.

But in reality many of these seeming truisms do not stand up to scrutiny. Despite fewer of us doing manual work, those who still do are not blessed with toned physiques and pleasingly low BMIs. In fact, 2013 data from the Health Survey for England showed that the highest rates of obesity were found in unskilled manual workers, at around 30 per cent for men and 35 per cent for women.[1] In professional classes, mostly performing sedentary office jobs, rates were around 20 per cent for men and 15 per cent for women. Although there are likely to be many confounders, not least diet and out-of-work physical activity, the fact that manual workers are twice as likely to be obese seems at odds with any idea that physical labour protected us in the past.

When it comes to how much exercise we actually do these days, there is very little reliable data. Adult engagement in organised sporting activity and membership of gyms has certainly increased as we have become more affluent, but exactly how many calories we are burning is near impossible to judge. And although organised physical activity in schools has decreased, attendance in out-of-school sports clubs has soared. And surprisingly, when the link between school physical education and BMI has been studied, the associations are incredibly weak. Although it does have a great many proven benefits beyond weight loss, particularly when it comes to physical confidence and coordination, school sport has not been shown to contribute to the prevention of weight gain or the reduction of obesity levels in any way.[2]

It may prove impossible to ever know how changing rates of physical activity have influenced the development of obesity up to now as we will never be able to determine how many calories were being burned fifty years ago. But what we can do is look at how different rates of exercise influence body weight today. Even if it does not quite fit with the calorie in – calorie out model as many might expect, we can say that exercise is very important indeed.

YOUR BODY WANTS COMPENSATION

If how fat we are is determined by the difference between how many calories we eat and how many we burn, then surely actively burning more should make us lose weight. Strangely, this does not always seem to be the case, but we probably all intuitively understand the reason why. Exercise is a really good way of building up an appetite, and whenever we have been active, we usually end up eating more. And although this might have an element of psychology to it, with many of us feeling we deserve a treat after going to the gym, it is also very much controlled by our hormones.

Many experiments have shown that after we exercise we are driven to eat more if we have free access to food. Experiments on military cadets have shown that these increases largely occur around two days after significant exercise, suggesting that it is not entirely due to the effect of a post-workout treat. Although the mechanisms are not entirely clear, our muscles produce a hormone called interleukin-6 during exercise, and this has a role in regulating our appetite.[3] And in experiments where rats are trained to run on tiny rodent treadmills, the gene that codes for leptin has been shown to be downregulated after their exertions, suggesting that this powerful hormone is also involved in post-exercise appetite control.[4] These sorts of processes help to ensure that no weight is lost in response to any increases in activity. Our bodies, as we have seen previously, really don't like losing weight and will powerfully respond to any changes in the environment that might make this happen.

Although the mechanisms of this regulation are probably many and varied, the control they exert is extremely precise. It seems that the number of extra calories taken on after exercise closely matches the amount expended, particularly for people engaging in high levels.[5] This would suggest that without accurate calorie control and the ability to resist hunger over several days, exercise is not a good strategy for weight loss, though that doesn't mean it's not good for your health.

Even though appetite control after exercise is extremely accurate, unfortunately it does not return the favour when we overeat. As most of us will know, physical activity does not go up after a huge takeaway, even when observed over long timescales. So when we eat too much, our bodies are likely to ensure that any excess calories are stored, rather than encouraging us to run around and burn them off.

To make things worse, Ancel Keys' Minnesota starvation experiments showed that when calorie intake is reduced, physical activity decreases. This is just one of the irresistible effects of our hunger hormones. So when we eat less, we will be naturally inclined to run a little slower, take a sneaky short cut, drive rather than walk, or give a little less on the football pitch. As usual, our body fights weight loss in any way it can. And the less you eat, the harder it will fight.

Perhaps worst of all, when we decrease our physical activity, we do not seem to compensate by dropping our calorie intake. So if you are currently exercising regularly and suddenly stop for some reason, your appetite and habits will cause you to eat the same number of calories that you always have. When this happens, most of the excess you take in will be stored as fat. When regular runners cut their distance down for a period of time, they quickly gain weight. And even when they return to running their original distances, the extra pounds often fail to shift.[6] This is one of the reasons why many professional boxers end up going to seed so quickly after retiring – it is an unavoidable consequence of our biology.

So, if you're looking for immediate effects, the depressing truth about exercise is this:

- If you are fat, exercising will not automatically cause weight loss, because you will have a strong urge to eat more.
- If you eat too much, your body will have no urge to burn off the extra calories.
- If you eat too little, you will automatically start to move less to prevent weight loss.
- If you exercise regularly, you'd better not stop, because if you do you'll end up getting fat.

CALORIES OUT

Anyone who has ever run on a treadmill will know another depressing truth about exercise. Even when it really hurts, it does not burn off that much energy. Depending on your weight and speed, a half hour run will probably burn 300–400 calories. And yet a 48-gram Snickers bar contains around 250 calories. So it does not take much delicious, convenient snacking to completely negate any impact of that hard-fought run, especially when the calorie compensation occurs over days rather than hours. This has led many, including the low carbohydrate diet advocates Aseem Malhotra and Tim Noakes, to claim that 'you can't outrun a bad diet'.[7]

But in reality, things are not quite so simple. Dr David Nunan is a senior research fellow at Oxford University who studies the role of physical activity in the prevention of lifestyle-related health conditions. He told me:

> You do have to do a lot of movement to use up 500 calories, and the headline grabber is that there is no benefit of exercise. Studies show that twenty minutes of daily exercise have no effect on visceral fat, but most of these are based on a version of walking. The studies look at levels of exercise that are within reach for people, but generally it is not enough. The conclusion from these studies should be 'regular people are unlikely to lose weight with a typical amount of a common type of exercise'. And that's if folk actually did the exercise they said they did. Supervised studies work better because it's more likely the exercise will actually get done but the reality is that getting people who have never exercised before to do enough, or hard enough, consistently, is difficult to say the least. But even the little exercise they do can be of some benefit. The majority of studies show that when it comes to weight loss, diet and exercise beats diet alone. It's just that diet has a bigger effect.

This view might seem at odds with the evidence that we eat more to compensate for extra calories burned during exercise. But this is perhaps because we only think about weight gain in terms of calories in and out. When it comes to exercise, although it might not help us get thin in the short term, there is good evidence that over longer periods, it can make a huge difference.

THE BIGGEST REGAINER

When Kevin Hall's team followed up on the *Biggest Loser* contestants six years after the competition, all of them had regained a lot of the weight they had lost. There was, however, a good deal of variability regarding how much had returned. Some were heavier than they had been before enrolling in the show, but others had managed to stay a good deal below their pre-show weights.

Perhaps strangely, the amount that people were eating seemed to have little to do with how much weight they had regained. The calories being consumed were nearly identical between those who had become heavier and those who had kept some of the weight off. The biggest difference between the two groups was how physically active they had become. Those who had managed to keep up a significant level of exercise seemed to have an advantage over those who did not, suggesting that exercise was the key to their long-term weight loss.[8]

Clearly, this was a small study and only looked at people who had been put through extreme weight loss treatments. But similar results have been seen elsewhere. There is in fact a great deal of research that seems to show that, although exercise has a limited role in initial weight loss, it is extremely important in preventing weight regain.[9]

There are many potential reasons why this might be the case. Exercise programmes have been shown to modify long-term regulation of appetite hormones, helping to increase perceived fullness after meals and making weight maintenance much easier. There is a suggestion that increasing physical activity might be one of the few things that can help to alter our bodies' set point weight downwards, so hugely increasing the chance that long-term weight maintenance might be possible.[10] If this is true, engagement in regular exercise could help free people from the misery of constant dietary failure and the endless cycle of yo-yo dieting.

For this reason alone, dismissing the role of exercise when it comes to weight loss, or claiming that 'you can't outrun a bad diet', is a dangerous and thoroughly irresponsible message. It will only encourage people to abandon exercise as futile, pushing them to engage in more and more restrictive diets.

Because exercise has little effect on initial weight loss, its importance is frequently overlooked. But the truth is, almost anyone can lose weight in the short term. That is the easy part and only requires people to eat a bit less for a while. It is keeping any lost weight from returning that is so difficult, and exercise seems to be the one thing that can narrow your odds.

Dismissing the importance of exercise is also likely to exclude people from one of the most powerful and effective lifestyle

interventions that we know of. Exercise has many extraordinary benefits beyond its effect on the body's set point weight. It can reduce the
odds of developing cardiovascular disease and type 2 diabetes, and can
improve many aspects of mental health, including some that increase
the risk of obesity.[11]

When people exercise while dieting, it improves their body composition, with more muscle and less fat under the skin. It also seems that
exercise is better at reducing visceral fat than diet. As we have already
discussed, visceral fat is thought to be a much more powerful predictor
of morbidity than BMI.[12]

This all suggests that although the effect of exercise on immediate
weight loss might be limited, it is actually far better than diet at improving our health. And for all of the many benefits, evidence seems to
indicate that the more exercise people can manage, the greater and
more positive the changes they are likely to see. And for those with the
most severe cases of obesity, it actually seems that the benefits of exercise can be even greater, reducing many of the risks of excess weight.[13]

Because of the way we confuse fatness, physical appearance and
wellness, exercise is too often dismissed. Weight is seen as the only
measure of health, when in reality it is a poor indicator. And so exercise
is frequently cast as a pointless waste of time. When dieters see that it
has no immediate effect on the scales, they assume it is not helping. In
our obsession with shifting pounds, we are ignoring one of the most
important things we can do to improve our health.

All the evidence shows that exercise is unlikely to help people lose
weight in the short term. But if they want to keep the weight off and
vastly improve their health while doing so, it is completely indispens-
able.[14] If we are serious about helping people lead healthier lives and
bringing down the prevalence of lifestyle diseases, exercise is perhaps
the most powerful and effective tool that we have.

There's just one problem. We're really shit at doing it.

THE EXERCISE PROBLEM

I am certainly not the first person to sing the praises of exercise. In
fact, although some dismiss its role in weight management, I still
haven't come across anyone who thinks its effects on general health
are overrated. If a new drug, diet or 'superfood' had a fraction of the

proven benefits of exercise, it would be hailed as miraculous, and fly off the shelves. Yet despite exercise being free, with profound and life-changing outcomes, very few people do enough. And in constantly being told they are not doing enough, they probably feel really guilty. (Sorry about that.)

Weight management guidelines often suggest that people who need to lose large amounts should engage in around sixty minutes of exercise daily. But studies have shown that the number of obese people who actually manage to achieve that as part of a weight management programme is virtually zero.[15] It appears that the recommendations are impossible for any real-life human being to achieve, which might lead us to question their usefulness.

Of course, you may think that fat people just need willpower. But can it really be the case that every single one of them is too weak-willed to comply? If you are thin, do you really consider every fat person you have met to have less willpower than you do? To be honest, if you still think that way, congratulations for getting this far into the book.

Presuming we can safely reject the idea that all fat people are lazy, this lack of compliance might indicate that something else must be going on. Could it be that the same hormone pathways that drive people to overeat also reduce their ability to perform significant exercise in a long-term, sustainable way?

It certainly seems that this is, at least in part, the case. Obesity and sedentary behaviour are strongly and consistently linked.[16] Remember how powerful the hormonal drivers are, nearly impossible to resist in the long term. These pathways do not just drive appetite, they also regulate every aspect of our energy expenditure, a significant part of which is exercise. Perhaps obese people have an inbuilt disadvantage when it comes to getting moving, the power of which the rest of us cannot imagine. And maybe the success of those people who manage to exercise and keep the weight off is because they have weaker hormonal drivers. Perhaps the ability to exercise is an effect experienced by people who find weight loss easier, rather than a cause of them being able to do so.

In reality, these effects are thought to be bi-directional, and likely to vary between individuals. So for some people, exercise will help them lose weight more easily. But others will be naturally more inclined to lose weight, while also finding exercise easier to maintain.

And there are other reasons why overweight people might struggle to exercise. Gyms, dance classes, running tracks, football pitches and tennis courts are often places where fat people are not made to feel comfortable or welcome. In researching for this book I collected many stories about weight discrimination. Some of the most unpleasant and depressing were connected to exercise and activity. I heard about people being openly laughed at and abused while running or cycling. Grown men reduced to tears, hiding in fear of looking ridiculous in newly purchased sports gear. Confident and high-achieving women told me how they were terrified of walking into a gym class, hiding away in their car for an hour before returning home in shame.

Frequently, overweight and obese people are told that exercise is not for them, even when they can achieve a great deal. I was contacted by a forty-five-year-old former army officer who had gained a large amount of weight as a young man after being injured during military service. He never managed to shift all the weight he had gained but still remained highly active. He told me:

> Two years ago I decided I was going to do whatever I could to get fit and lose some weight. I dropped from about 130 kilo-grams to 110 kilograms and started doing Ultra Marathons. One of them was called Thunder Run, a 24-hour race where the person that goes the furthest wins. I managed 100km and some-how everyone thought that this was extraordinary. Eventually, I realised that what they meant was, it was extraordinary for a fat guy. I also managed eight half marathons, a full marathon and three trail half marathons. I think the most annoying thing was that after each race, you would line up to get your commemora-tive medal and t-shirt. For the most part, the largest size they would have is an XL, or on rare occasions an XXL. But being sportswear, an XXL is the equivalent of a large in any other product, so in nearly one hundred races, I never had a race t-shirt that fitted.

This is certainly not an unusual or isolated experience. Studies of frequent exercisers, exercise professionals and personal trainers have shown the existence of high levels of weight prejudice within these communities. Surveys of regular exercisers have found widely held beliefs that body weight is controllable, obesity is caused by poor choices and fat people lack willpower. They also openly consider anyone overweight to be ugly, lazy and gluttonous.[17] And yet we expect fat people to happily enter into that environment, filled with prejudice and demeaning judgement. And then we demand they exercise for sixty minutes a day, which very few thin people ever manage, only to tell them they are failures when the pounds don't drop off.

Dr Deb Burgard is a psychologist based in California and one of the founders of the Health at Every Size movement. She has spent many years teaching dance classes for large women, as well as working with eating disorder patients. She describes her classes as being 'not a weight loss class, nor a penance', simply designed to encourage people with few positive experiences of exercise to move their bodies in a safe environment. The classes have proved extremely popular over the years, with available spaces few and far between. She told me:

> Some people have good memories of physical activity, especially at recess/break, but for others it was traumatising. If the playground was a site of teasing or bullying, or made the person feel uncomfortable or unskilled or like they didn't belong, physical activity can trigger those experiences. We all have a tendency to blame our own bodies for the meanness of other people, rather than directing our anger at the meanness. Later in life, when people – at any size – are trying to move their bodies or learn a skill, it can feel like it is not safe to try things out. And that is just the barrier from the inside – there are all sorts of barriers in the world as well.

Dr Burgard's classes attempt to create a safe space for larger people to experience and enjoy movement, free from stigmatisation and shame. But in many ways, it is saddening that a class like this is necessary. She explains:

> Thin people might think, 'Higher weight people don't come to the gym', but the problem is located in the gym. Imagine the barrier of trying to enter a space that is designed to eliminate people like you, filled with people who fear looking like you, where the assumption is that you are there because you are ashamed of your body. No, just no. When an environment is respectful and welcoming, it makes a huge difference.

Environments where higher weight people are trainers and instructors, where the equipment and activities are geared to higher weight people who are moving more mass just by being active, provide a chance to build the same cardiovascular conditioning, strength and flexibility as anyone else. We really need to start designing activities for the full range of people's bodies, rather than expecting them to change their bodies to fit the activity. We need to create environments where bodies are not mocked or ranked, where the different gifts of different bodies can be celebrated, and everyone can participate. And mostly, we need environments where all people can experience either greater purpose or greater joy, or both, from the movement they are doing, so they want to keep doing it.

Stigma and prejudice are certainly significant factors driving fat people away from exercise environments. It is completely unacceptable that this sort of prejudice should exist anywhere. But when it is in a place where such huge health benefits lie, it seems all the more perverse and destructive. A privileged thin stride around their tower of exercise and activity, ridiculing fat people both for being who they are and for trying to change. If this shameful admonishment did not exist, just think how many people might improve their health. Just imagine how much better our world might be.

LAST WORD

A final word on exercise. It has extraordinary benefits. It can be a source of real pleasure and greatly enrich our lives. For some, it brings them together in shared goals. It can help form new friendships, and cement existing bonds. For others, like myself, it can provide an oasis of quiet reflection in our busy and stressful lives.

If you move your body regularly and get pleasure from it, be happy you have found that space. It has more benefits than anything else we can do in our lives and is a million times healthier than any restrictive weight loss diet.

But we must also try to accept that exercise will not be for everyone. Some will never be able to find a type of physical activity they can enjoy or maintain. Willpower alone will not keep you at the gym, and if you cannot find pleasure in movement, then it is unlikely you will last at it for long. For others, the stigma and traumatic associations will be too powerful, and it will not be worth the stress that these create. Injuries or disability might make all but the most tentative movement dangerous and painful.

So although I would strongly encourage everyone to get as much movement into their lives as they can, I appreciate that for some this will not be possible. We all need to try to lead the best lives we can, and no one should be made to feel guilty about things they cannot do. Because although a lack of exercise might cause you some problems down the line, I guarantee that feeling guilty about it will fuck you up twice as quickly.

12

IS IT BECAUSE OF FAT?

It ain't what you don't know that gets you into trouble. It's what you know for sure that just ain't so.

Mark Twain

We are drawn to essentialism when it comes to food. We imagine, quite wrongly, that we take on the properties of whatever we eat. The psychologist Paul Rozin once observed that when people are told of a fictional tribe that hunts turtles for their meat and wild boar for their bristles, they imagine that they will be excellent swimmers. But when they are told that the same tribe hunts turtles for their shells and eats boar meat, they envisage them as fierce warriors. Although most of us realise that 'you are what you eat' is a ridiculous oversimplification of our complex metabolism, we keep thinking it anyway.

It is perhaps for this reason that we are so drawn to the idea that eating large quantities of fat will make us become that way. This is especially powerful when we picture fat from rotund farm animals such as mud wallowing pigs or lazy, slothful cows. These animals lounge around their fields all day without a care in the world before they are fattened up for the kill. Steaks and chops with rich layers of wobbling fat are the most prized and flavourful. Although uniquely delicious, when there is a lump of mammalian adipose tissue on the plate in front of us, it is hard not to imagine that it will transfer to our midriffs the moment it is consumed.

There is some justification for this. After all, fat is the most energy-dense foodstuff we consume, containing nine calories per gram. Sugar and other carbohydrates have only four calories per gram, and protein roughly the same. Fibre averages out at around two calories per gram for nutritional labelling purposes, although exactly how much is converted into energy will depend on which type of fibre it is, and on the efficiency of your microbiome. As anyone who has eaten sweetcorn will know, some fibre passes straight through unmetabolised.

So one might think that a diet with plenty of protein, carbohydrate and fibre, but relatively low levels of fat, would be a good way to go if you are planning on losing weight. Fat is an extremely efficient (and extremely edible) calorie delivery system. There is even some recent evidence that fat, once thought to provide nothing more than richness and texture, might have a unique taste all of its own. The nutrition scientist Professor Richard Mattes, whose work into beverage consumption we have already discussed, has coined the term 'oleogustus' to describe this unique taste, and proposed that it should be added to the existing palate along with salty, sweet, savoury (umami), bitter and sour.[1] He explains:

Triglycerides are tasteless, but free fatty acids have an unpleasant, rancid taste. Olive oil is graded, in part, by the taste of these free fatty acids. The taste is most probably a warning signal, like bitterness. Although few people would drink something that is just bitter, it can add complexity in the right context. In the same way, the taste of fatty acids may theoretically be a pleasant hedonic experience. Our work also shows that the taste of fat might be important for modulating fat metabolism, potentially modulating lipid stores in the intestine and triglyceride concentrations in the blood.

Fat is every chef's friend, and many people would be shocked by the amount added during much restaurant cooking, even in our health-conscious era. Ever wondered how chefs get their sauces so rich and indulgent? How they can make purées and glazes so flavoursome and irresistible? Often it is through generous use of butters, oils and animal fats. Perhaps the biggest shock, when I first started working in food manufacturing, was having the nutritional content of my recipes analysed for the first time. I discovered that many of my favourite dishes easily provided most of an adult's recommended daily amount of fat, and sometimes even more. I had to considerably readjust my techniques and expectations in order to achieve anything that could be sold on a supermarket shelf.

As well as having an extremely high-energy density, fat is also less likely to combine with water than starch and protein, meaning high-fat foods are often particularly dense in calories. High-fat items are also thought to be less satiating than many other foods, meaning that we are likely to become hungry again soon after eating. It is also thought that when we take on calories as fat, there is less resistance to the excess being stored, making it more likely to go straight to our adipose tissue. Perhaps because of this, when people are overfed on a high-fat diet, their body fat increases to a greater extent than when they are given the same number of calories as carbohydrate.[2]

When the amount of fat in a meal is covertly increased, it leads to people taking on more calories,[3] which, given that it tends to make things more delicious and energy dense, is perhaps unsurprising. In the US, as rates of obesity have risen, so has consumption of dietary fat, particularly through fatty convenience foods and indulgent baked goods. So the case against fat as a driver of obesity seems pretty tight, indicating that our essentialism might actually have some merit.

As well as weight gain, consumption of high-fat diets has also been implicated in the development of cardiovascular disease. Throughout the 1950s and '60s, as rates of heart attacks and strokes rose, the role of saturated fat and cholesterol came under increasing scrutiny. Huge increases in meat and dairy consumption seemed to be driving ill health. As this reached crisis levels in the 1970s, there were calls for something to be done.

New US dietary guidelines were put in place in the late 1970s, with the UK following suit shortly afterwards. There were calls to reduce saturated fat and cholesterol, and to increase consumption of fruit, vegetables and fibre. In response, food manufacturers and the diet industry focused on low fat foods, often replacing fat with sugar or carbohydrates. Reduced- or zero-fat options became common, and low fat plans started to dominate the diet industry. Fat became public enemy number one, and manufacturers and retailers bent over backwards to give people the products they craved.

The result was . . . Well, it really depends on your point of view. Many YouTube conspiracy theories consider these guidelines a disaster, as evidenced by rising rates of obesity. For now, it is worth remembering that although obesity rose rapidly around this time, the actual rate of BMI increase was only continuing as it had for the previous fifty years. At worst, the new guidelines had little effect on people's weight. There is certainly no evidence they produced a seismic shift.

When it comes to the risk of CVD, however, there were significant changes. It is nearly impossible to assess how much of the change has been due to any specific effects of fat in the diet, but the rates did fall, and they continue to do so.[4] Although heart disease is still a significant killer, we have made great progress.

It is also worthy of note that although similar dietary guidelines were introduced in many countries, there was actually very little in the way of consistent change when it comes to what people were eating. In the US and Canada, consumption of fat increased. In the UK it decreased, both in total and as a percentage of all calories consumed. The one thing everyone knows about dietary guidelines is that nobody follows them, so assigning any special significance to the changes seems a little silly.

THERE'S FAT AND THERE'S FAT

It would seem that the case against dietary fat is pretty open and shut. Fat has loads of calories, it is delicious, and it makes you fatter than other foods when you eat the same amount. Cutting back would seem to be an excellent weight-loss strategy.

So why are we even discussing it? Well, somehow dietary fat has become one of the most divisive and confrontational subjects in modern nutrition science. Although many will cheerfully accept the evidence I have set out in the introduction to this chapter, and think we might benefit from cutting down, others will be spitting blood, furious that I have grossly misrepresented the science. Fat, and the demonisation of it by generations of nutritional scientists, divides people into religion-like sects. Some believe that fat is a dietary evil making us sick. Others swear that embracing fat and getting more into our diet is the key to reversing the obesity epidemic.

So who is right?

Perhaps much of the confusion stems from the fact that all fats are not created equal. Even though they have a similar number of calories per gram, different types can affect our bodies in different ways. Some are absolutely essential for our functioning, and it would be extremely unwise to try to cut all fat from our diets. Crucially, fat is not just an energy store. It serves a number of important functions in the body, including roles in the creation of cell membranes and the production of hormones.

There are many types of fat, and a good deal of complexity about how they function in the body. This complexity is at the heart of much confusion and debate, largely because of some catastrophic mistakes made regarding dietary-fat recommendations in the past. This has led to mistrust in the nutritional sciences, calls for all guidelines to be rejected and the demonisation of the food-manufacturing industry. To understand how this happened we need to understand a bit about the chemistry.

The fats in our diet are composed of triglycerides, consisting of a glycerol head, with three long chains of carbon atoms attached. Most of the carbon atoms along these chains are attached to two hydrogens. If all of the carbons have two hydrogens attached (apart from the one at the end which has three), then the fat is described as being saturated. This means that all the carbons are saturated with the maximum possible number of hydrogens. If there are a couple of hydrogens missing somewhere along the chain, then two carbons next to each other can form a double bond between them. When this happens, you have a monounsaturated fat. If there is more than one double bond along the chain, then you have a polyunsaturated fat.

Saturated Fatty Acid

Unsaturated Fatty Acid

Polyunsaturated Fatty Acid

It is thought that large amounts of saturated fat in our diet can increase our risk of cardiovascular disease. Indeed, systematic reviews of all the evidence show that replacing some saturated fats with polyunsaturated fats in your diet is likely to have a positive effect on health, explaining why this is a common recommendation from different authorities.[5]

Even this is hotly contested by many people, but the quality of evidence seems pretty convincing to most. The general advice is to cut down on foods high in saturated fat, such as red meat, butter and coconut oil, and replace them with olive oil, oily fish, nuts and seeds. Although red meat and butter are delicious, it seems that in large quantities they are not ideal.

THE BIGGEST MISTAKE IN FOOD

Because they are largely produced from animal products, sources of saturated fat tend to be considerably more expensive and likely to spoil more quickly than unsaturated fats. Unfortunately you can't just swap one for the other – they don't taste the same, and they don't behave the same way in spreads and baked goods.

For these reasons, and because the burgeoning market for soy protein was producing a lot of spare unsaturated soy bean oil, in the early part of the twentieth century industrial processes were developed to turn unsaturated fats into saturated ones. A process called hydrogenation adds the missing hydrogens where the double bonds are, giving vegetable-based fats different textures and properties. Saturated fats tend to be more solid at room temperature, and with the level of control that chemical hydrogenation brings, the exact melting points and solidity can be carefully controlled.

As this technology was developed, it led to a raft of new brands and products, with the first vegetable-based cooking fat launching in 1910 under the brand name Crisco. Because of the versatility and cheapness, hydrogenated fats quickly dominated the market for shortenings and spreads. And as refrigeration started to become more commonplace in homes, the ability to produce products that could be used straight from the fridge gave these oils a unique advantage over butter, lard and dripping.

To be clear, saturated fats produced by hydrogenation are exactly the same as those made naturally. But when partial hydrogenation occurs, where the missing hydrogens are not entirely replaced, fats with an unusual structure can be formed. Whereas most naturally occurring unsaturated fats have a type of double bond called a 'cis', partial hydrogenation produces fats that have something called a 'trans' bond, with a slightly altered shape and differing metabolic properties. And these 'trans fats' were to become something of a problem.

Cis vs Trans Bond

Cis Configuration   Trans Configuration

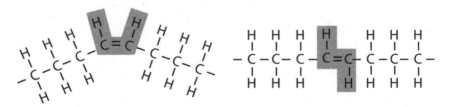

Trans fats are not an entirely artificial creation, and are naturally found in the meat and dairy products of ruminant animals. At low levels, they are even found in human breast milk. The trans bond is not produced naturally by mammalian metabolisms, and instead is a product of our gut bacteria. Butter naturally contains around 3 to 4 per cent trans fats, meaning that we have long been exposed to them in our diet. The difference with partially hydrogenated oils, however, is the amount of trans fat contained within them. Some vegetable shortenings contain around 30 per cent, and some margarines around 15 per cent. Widespread consumption of these partially hydrogenated oils meant that many people were taking on more trans fats than at any time in history.

Ironically, the success of partially hydrogenated spreads received a boost in the 1960s, as evidence for the cardiovascular harms of saturated fats came to light. People turned to them in preference to butter and lard, because they contained lower levels of saturated fat, and more of the supposedly healthier unsaturated type. It was to be many years before the metabolic differences between the trans and cis unsaturated fats were fully appreciated, and by then, much damage had been done.

It turns out that trans fats have a uniquely harmful effect on the body, thought to be due to them interfering with the metabolism of essential fatty acids, and altering the composition of our cell membranes. This leads to inflammation, contributing significantly to a number of health problems. By the mid-1990s, the extent of the damage caused by trans fats was starting to be revealed. Consumption had been shown to raise levels of LDL cholesterol (the so-called bad cholesterol), lower HDL ('good') cholesterol, and increase the levels of triglycerides in the blood. They had also been strongly linked to the development of CVD, type 2 diabetes and obesity.

A 1994 report by the Harvard nutrition professor and firebrand Walter Willett suggested that trans fats from hydrogenated vegetable oils (HVOs) were responsible for a quite staggering 30,000 coronary heart-disease fatalities annually. The worldwide death toll since their creation in 1910 is too terrifying to contemplate.[6][7]

Hydrogenated vegetable oils are one of the great failures of modern food manufacturing. But thankfully the reaction to new evidence was swift and decisive. Although a shameful 1995 industry-sponsored report suggested that evidence was not sufficient to take any serious action,[8] very soon pressure to remove HVOs from our food supply became overwhelming. Products were quickly reformulated, often using more complete hydrogenation products, combined with conventionally unsaturated cis fats. Soon, the majority of vegetable-based spreads and shortenings had fewer trans fats than their animal counterparts, and levels in the food supply had dropped below what is generally considered harmful.

The World Health Organization recommends that trans fats should provide less than 1 per cent of a person's calories, and in the UK the average consumption is comfortably below this. In the US, at the time of writing, partially hydrogenated oils have been almost completely removed from the food supply, and a total ban came into force in May 2018. In Denmark, a complete ban on partially hydrogenated vegetable oils led to a 50 per cent reduction in deaths from coronary heart disease within just twenty years.[9]

That is not to say that we should be complacent. In the UK, the Food Standards Agency has resisted an outright ban, favouring a voluntary approach from the food industry. Although this has been very successful in reducing average consumption, partially hydrogenated oils are still used by a number of independent fried chicken outlets, meaning that people who like their wings are at risk of consuming dangerous levels. Perhaps for cultural reasons within these establishments, combined with a false perception that the extra shelf life of HVOs offers better value for money, the use of these deadly oils persists. As the number of people who eat regularly at these shops are small compared to the population, their consumption is not enough to affect average figures, but it could be devastating for the health of these individuals.

There is no reason for shops to use these oils, as a number of safer, cost effective alternatives are available. One can only assume that shop owners are ignorant of the potential for harm. It is utterly shameful that the manufacturers and wholesalers, who definitely know better, still allow HVOs to be sold. It is also deeply disappointing that in the UK there are currently no plans for legislation to protect people. This inaction may be due to a certain contempt for those who frequent chicken shops, often poor and often not white, who are deemed to be irresponsible and not worth bothering with. We have finally found a simple intervention that will definitely help people, without inconveniencing them or limiting their choices – and our response is to do nothing.

Apart from these few cases, it appears that consumption of trans fats, potentially a factor in increasing rates of obesity, is now very much under control. It should be noted, however, that this is only the case in a limited number of developed countries. In some parts of the world trans fats are still common, and in some areas becoming more popular. In South Asia particularly, HVOs are increasingly being used to replace ghee and shortening for the making of traditional food items such as samosas, parathas and pakoras.[10] Consumption is also thought to be high in a number of Middle Eastern countries: it has been estimated that in Iran around 39 per cent of all heart-disease deaths could be prevented if HVOs were removed from the food supply.

It would be a shame to make such big progress in one part of the world, and not let others learn from it. Partially hydrogenated vegetable oils are not consumed as part of anyone's culture. They are not an inevitable consequence of poverty. They are eaten through ignorance alone, a lack of knowledge of the harm they do, and of the many more healthful options available to replace them. If we are serious about tackling obesity worldwide, a global approach to HVOs would be a good place to start.

CHOLESTEROL

Perhaps the second great misstep that nutrition science made when it comes to fat regards cholesterol. In 1984, *Time* magazine carried the headline 'Cholesterol is Proved Deadly – and our diet may never be the same again' in response to a major US government investigation into the effect of blood cholesterol on the development of heart disease. The project director of the study, Basil Rifkind, was quoted as saying that 'the more you lower cholesterol and fat in your diet, the more you reduce your risk of heart disease'. This news created a seismic shift in the advice being given out about food and health.

The *Time* article suggested that you would 'never look at an egg or a steak in the same way again'. There was a particular focus on eggs, something that until then had been thought of as wholesome and nutrient packed, because egg yolks contain large quantities of choles-terol. Egg consumption was strongly implicated as a major cause of CVD, and along with other cholesterol rich foods they were demon-ised as being profoundly unhealthy.

Looking back, this was a great shame for the humble egg, espe-cially given what a cost-effective and widely eaten food it was. Some go as far as suggesting that the demonisation of eggs led to significant shifts in dietary patterns, particularly when it came to breakfast. As foods containing supposedly artery clogging cholesterol were roundly vilified, breakfast cereals rose to prominence, often full of sugar and starchy carbohydrates. In shaming one food as a dietary evil, many think that a greater problem was created, as people lurched from one unbalanced diet to another. And to make matters worse, it turned out that cholesterol was not quite the villain it had been made out to be.

It has long been known that cholesterol is a hugely important molecule, with many vital functions in the body. It helps form cell membranes, regulates hormones, and is heavily involved in digestion. It is true that aberrant blood cholesterol levels are linked to people's risk of CVD, but the implication that this was largely driven by cholesterol in people's diets was sadly mistaken. It was a leap taken by researchers and nutritional authorities at the time, partly because it seemed obvious that you are what you eat.

At this stage it would be great to say that dietary cholesterol has been completely vindicated, and we can all eat a carton of eggs a day. That would be the simple message that everyone wants to hear, with eggs going from zero all the way back to hero, and being carried (carefully) through town to a tickertape parade. In reality, the extent to which dietary cholesterol affects blood cholesterol varies from person to person, dependent upon how good our body is at regulating production in response to intake. For most people, when we eat foods high in cholesterol, our body quickly shuts down its own production, regulating how much is circulating in the bloodstream. But for others with less efficient control mechanisms, too many eggs might still cause a problem. This could be a matter of genetics, lifestyle, disease response, or perhaps something else. As is often the case with nutrition science, the closer we look, the more there is to discover. Anyone offering simple answers is probably trying to sell you something.

So, although the relationship between dietary cholesterol and cardiovascular disease has probably been over-egged (sorry), for some there might still be a risk.[11] That said, a Harvard egg study conducted in 1999 indicated that eating seven eggs a week did not seem to carry any increased health risks.[12] *

It is probably true to say that dietary cholesterol has largely been exonerated. However, high levels of saturated fat in the diet are still thought to have a negative impact on blood cholesterol and increase the risk of CVD. Although the debate rages on, the health benefits of cutting down on saturated fat probably depend on what you replace it with, the preferred option being polyunsaturated fats.[13]

And no one with any sense is saying that we should stop eating

---

* Well, sort of. In male patients with diabetes, the risk of CVD went up, although the reasons why are still unclear.

saturated fat completely, particularly me. I am a lifelong advocate of butter and cream, and consider the joy that they bring to be worth a little risk. Anyone with a sensible attitude to food would not want to eat these things at every meal. But I suspect that when eaten occasionally, the pleasure and positivity that good cooking can provide are enough to counter the harm of the odd butter sauce.

Despite sometimes falling into the trap of using them myself, I loathe the terms 'good fats' and 'healthy fats', because they imply that other fats are 'bad' or 'unhealthy', which just is not the case (apart from trans fats obviously, but you know what I mean). The healthfulness of a food depends far more on what it means in your life than on the particular mix of nutrients it contains. So although I err on the side of suggesting we should cut down on saturated fat, I would never want anyone to feel the slightest bit guilty for eating bacon, butter, steak and cheese (which, incidentally, would make an amazing sandwich).

In truth, the whole fat debate is full of nuance, and the best that anyone can say is that different people probably react very differently to different diets. Some might flourish on foods packed full of saturated fat and cholesterol, where others might struggle, desperately craving carbs. This might be determined partly by your genes, but probably also your life stage, meaning that what you eat as a young woman might not be suitable as you age.

For most people, the best recommendation is to eat fats from a wide variety of different sources, as this will ensure you are not getting too much saturated fat, are not eating excessive cholesterol, and are taking on plenty of unsaturated fat. If you spread all the different potential sources of dietary fats out in front of you, there are far more unsaturated ones. All can be delicious in the right context, so if you really love food, you will not want to miss any of them out. So as usual, unless there are medical reasons for you to be following a particular diet, you should eat a little bit of everything, mix things up as much as you can and, most importantly, try not to worry. You'll probably be fine.

The truth is, the moment we start favouring one particular food and demonising another, we seem to get ourselves in trouble. We should not think of anything as the devil, nor should we idolise specific foods and consider them essential.

Well, actually, that's not entirely true . . .

THE GREAT OMEGA

Although we can probably take or leave most dietary fats, it turns out that some actually are essential. It is important here to distinguish between the common usage of the word 'essential' and the way it is used in nutrition science. For most of us, 'essential' means something really, really important. 'Follow these ten essential tips for perfect abs', '*The Angry Chef* book is this season's essential, must-have accessory', 'It is essential that I watch the new episode of *Westworld* this evening'. But when it comes to nutrition, essential means something that we need to take on through our diet because our body cannot make it.

When it comes to fats, our body is pretty good at making most of them, sometimes a little too good. It is particularly efficient at making cholesterol, presumably because this performs so many vital functions, and we don't want to be in any danger of running out. We can only get dietary cholesterol from animal sources, and although we are an omnivorous species, it is likely that throughout our evolutionary history there would have been times when we would have been forced into consuming a plant-only diet. In fact, if plant foods were abundant at particular times of the year, it is not hard to imagine that early humans might have foregone the dangerous activity of hunting in favour of some literal low-hanging fruit. If humans had not had the ability to make cholesterol, they would have quickly become sick if they went without meat or eggs, leaving them extremely vulnerable to changes in the environment. So it is unsurprising that cholesterol is not one of our essential nutrients, allowing us to live largely off plants if we really need to.

We are a remarkably versatile species, capable of adapting to many different diets, and so there are very few nutrients considered essential. These are mostly vitamins, minerals and some of the amino acids that form the building blocks of proteins. All but a very few are available from both plant and animal sources. It is important to consume all of these on a regular basis, which is why a wide and varied diet is the best option. Within fats, although we can make most of them, there are two particular ones that are considered essential. These are known as Omega-3 and Omega-6 fatty acids. They are both unsaturated, with the Omega number defined by the position along the chain where the double bonded carbon sits. Omega-3 fatty acids have a double bond on the third carbon from the end, Omega-6 have one on the sixth.

As our bodies cannot make Omega-3 or Omega-6 fats, it is extremely important to get them into our diets in sufficient quantities. They are building blocks of our cell membranes, and particularly important in the development of our brains and nervous systems. Deficiencies can lead to a number of problems, particularly neurological and optic issues in children, although it is generally thought that intake has to be extremely low before this occurs.

THE 3 TO 6 RATIO

Actually, I have already oversimplified somewhat, and nutrition nerds around the world will be cross with me. In fact, there are several types of Omega-3 and Omega-6 fatty acids, but only two specific ones that are considered essential. This is a fairly technical point, but it is worth a bit of explanation, because many consider that within these two essential fats lies one of the reasons why obesity has been rising. And to properly understand why these people are wrong, we need an explanation of why not all Omegas are equal.

It is only the two simplest members of the Omega-3 and 6 families that cannot be synthesised, namely Linoleic Acid (LA, Omega-6) and Alpha-Linoleic Acid (ALA, Omega-3). Linoleic Acid is commonly found in eggs and many vegetable and seed oils, including corn oil, sunflower oil and soybean oil. Alpha-Linoleic Acid is found in green plant tissue, rapeseed oil and certain nuts and seeds. The body can make the other Omega-3 and 6 fatty acids as well as obtaining them from dietary sources, so they are considered non-essential. Interestingly, the Omega-3 oils that we get from oily fish are actually some of these non-essential forms. But it's worth knowing that making our own non-essential Omega molecules still requires us to eat LA or ALA, which act as building blocks.

In recent years, a number of people have suggested that the ratio of Omega-3 to Omega-6 in our diet might be important for our health, particularly in the development of obesity and metabolic disorders. Some have gone as far as suggesting that the change in this ratio over time is one of the key drivers of metabolic problems, and is placing a huge health burden on human populations. Here are some of the more interesting approaches:

If you want to increase your overall health and energy level, and prevent health conditions like heart disease, cancer, depression and Alzheimer's, rheumatoid arthritis, diabetes, ulcerative colitis, and a host of other diseases, one of the most important strategies at your disposal is to increase your intake of omega-3 fats and reduce your intake of **processed** omega-6 fats.

Joe Mercola, yet again

The increase in the intake of refined seed oils and the concomitant reduction in the intake of omega-3 fats . . . is the most dramatic dietary change that has occurred in the last 100 years. Never before in human history have we ever consumed these seed oils and they may be increasing the risk of chronic disease including obesity.

James DiNicolantonio, discussing a controversial paper he published on the subject in 2016

In 2017, UK cardiologist Aseem Malhotra and former athlete Donal O'Neill published a ridiculous book called *The Pioppi Diet*, where they sing the praises of southern Italian cuisine, before providing some bizarre and very un-Italian recipes involving lots of coconut oil and 'pizza' bases made from ground-up cauliflower. Among their awful takes on one of the world's great cuisines, there was an explanation of why they think the Omega ratio is such a serious issue. They claim that an imbalance between Omega-3 and Omega-6 is the root cause of heart disease, colorectal cancer, breast cancer and rheumatoid arthritis. Showing remarkable insight into the eating habits of ancient humans, they claim that the ratio within the diet of our Palaeolithic ancestors was one to one, but the modern ratio is around twenty-five to one, causing us all sorts of ill health.

It is worth being extremely suspicious of any claims to have this level of insight into the exact diets of our ancient ancestors, largely because cavepeople selfishly failed to leave us any food diaries (I admit I made an educated guess about periods of vegetarianism earlier). Anthropologists who study the food eaten during the Palaeolithic have far less insight than most modern diet gurus seem to have, because there is very little archaeological evidence that might indicate food consumption patterns. But this does not stop the Omega ratio argument from being very popular in the so-called paleo-diet community, largely because it provides a little more pseudoscientific justification for the wholesale rejection of modernity. It also helps a whole plethora of low carb dieters patch some of the most obvious holes in the arguments for their diet.

KENTUCKY FRIED PROBLEM

Low carb, high fat dieting (LCHF) has become an extremely popular weight loss strategy in recent years, but it does have a problem. Many claim that it is the key to good cardiovascular health, and that carbohydrate consumption is uniquely linked to obesity and heart disease. We shall discuss this more in the next chapter, but for a diet that claims to be health promoting, it seems strange that it would seem to be endorsing the free consumption of items such as deep fried chicken. KFC is something very few people would endorse as part of a healthy lifestyle, but if the diet you are selling claims that low carb, high fat foods are good for you, then presumably, fried chicken is a perfect and delicious option. As long as you hold the fries.

So, instead of accepting there might be flaws in the premise of high-fat diets, advocates simply claim that the type of fat is all important, even when it is the supposedly healthier unsaturated type. Everyone now agrees trans fats are bad, but if you can expand this to include vegetable seed oils, you can still demonise most of the food that normal people enjoy eating, and explain why many people on high-fat diets seem to be unhealthy. This may be why the likes of Mercola, Malhotra and O'Neill, all LCHF diet enthusiasts, cheerfully embrace the Omega-ratio argument. It seems to patch up a big fried-chicken-shaped hole in their diet philosophies. Of course, if chicken is fried in an Omega-3-rich rapeseed oil, as much is these days, then the argument falls apart. But this is a detail that most low-carb enthusiasts choose to ignore.

The question is, why would this ratio matter anyway? Is there any justification for the paleo inspired claim that the ratio of Omega-6 to Omega-3 is the root of all our dietary problems?

The answer lies in how these fats are used in the construction of cell membranes. Everyone agrees that Omega-3 and Omega-6 oils are both extremely important, and we need both of them in our diet. But Omega-3 is also thought to have an anti-inflammatory role. If a cell membrane has lots of Omega-6 and not very much Omega-3, there will be a tendency for inflammation, which might lead to a number of metabolic diseases, including obesity. So if we eat more Omega-6 than Omega-3, does additional Omega-6 get incorporated into our cells? Because if it does, there might actually be a problem.

When Linoleic Acid (Omega-6) and Alpha-Linoleic Acid (Omega-3) are being used to make new cell structures, they both compete for the same enzyme, something called Omega-6-Desaturase. This enzyme turns these two small fatty acids into the larger, more complex versions that are used to make cell membranes. It has been suggested that if there are large amounts of Omega-6 around, they will dominate the enzyme, blocking Omega-3 and stopping it from being used. This means that if the ratio of Omega-6 to Omega-3 is high, too much Omega-6 will end up in the cell membranes, so causing inflammation and disease.

This makes a good deal of intuitive sense, and until relatively recently there was some serious scientific interest in the ratio and its effect on health. But it turns out that the total amount of ALA (Omega-3) in the diet is the important factor, rather than the ratio. So, if we eat enough ALA, it will be incorporated into our cells in the correct amounts, independent of any other Omega-6 fats we might consume. And population studies seem to bear this out, showing that diets high in LA actually seem to be associated with lower risks of cardiovascular disease, insulin sensitivity and inflammation.[14] [15] [16] [17] [18] [19]

Equally, the very idea that the ratio of Omega-6 to Omega-3 oils has recently increased does not stand up to scrutiny. Professor Tom Sanders, one of the UK's leading authorities on dietary fat, commenting specifically on a 2016 paper from Artemis Simopoulos and James DiNicolantonio, was quoted as saying:

The article wrongly suggests that the increase in the ratio of omega-6:omega-3 fatty acids in human diets coincides with the obesity epidemic. The ratio increased in the 1970s to 1980s but has either been relatively stable, or more likely has fallen over the past two decades, during which obesity has rocketed. Indeed, food manufacturers have actively sought to lower the ratio mainly because of emerging evidence regarding the health benefits of omega-3 fatty acids with regard to cardiovascular disease and visual function. This is very evident from the composition of high in polyunsaturated fatty acid margarine in the 1980s compared to nowadays, where the ratio has dropped from 131:1 to 4:1. There has also been increased use of new high oleic varieties of sunflower oil and soybean oil for deep fried foods as well as increased usage of rapeseed oil (Canola), which has a relatively low ratio.[20]

In addition to this, a number of experts have expressed annoyance that when looking at foods, the ratio that people talk about tends not to be the ratio of LA to ALA, but includes the non-essential Omega-3 and Omega-6 fatty acids, which are irrelevant to the discussion. These larger fatty acids do not compete for the Omega-6-Desaturase at all. So unless the ratio can actually be broken down to LA and ALA alone, it is not even of theoretical use. So anyone not breaking the dietary ratio down to these specific essential fatty acids is misunderstanding the science behind their own argument, something definitely worth reminding them of next time you meet your friendly local paleonutter.

Interestingly, there is also some debate in the literature about how much ALA in our diet gets converted into the more complex Omega-3 oils, with some researchers postulating that only about 1 per cent does. This would mean that ALA is less important in the diet than we think, as the vast majority of the complex Omega-3 fats used to build cell membranes comes directly from dietary sources, never competing for the Omega-6 Desaturase enzyme at all. As a result of this and other evidence, a 2007 UK Food Standards Agency report concluded that 'the Omega-6:Omega-3 fatty acid ratio is not a useful concept and that it distracts attention away from increasing absolute intakes of long-chain Omega-3 fatty acids'.[21]

In 2008, a paper on fat consumption produced for the Food and Agriculture Organization of the United Nations stated that, 'Based on both the scientific evidence and conceptual limitations, there is no compelling scientific rationale for the recommendation of a specific ratio of n-6 to n-3 fatty acids or LA to ALA, especially if intakes of Omega-6 and Omega-3 fats lie within the recommendations established in this report.'[22]

There is no question that getting plenty of Omega-3 into our diets is a good idea, and it also pays to get this from a variety of different sources. Even though in literal terms only ALA is essential, it does seem that the more complex forms, especially those in oily fish, are also extremely important. But again, getting enough is probably best achieved by eating a wide variety of different stuff, and not worrying too much. Fish is delicious, especially oily fish, so the comforting news is that if you enjoy eating nice things you are unlikely to have a prob-lem with the fatty acid composition of your cell membranes.

It has been known for over ten years that the Omega-3 to Omega-6 ratio is not a useful tool in assessing how healthy a diet is, but a number of fringe nutrition voices still seem to advocate it. Although this might seem harmless, it is not. The very best evidence for reducing the risk of cardiovascular disease suggests that replacing saturated fat with polyunsaturated fat is a good strategy, and increasing consumption of oils containing Omega-6 polyunsaturated fats is an excellent way of achieving this. Demonising cost-effective and palatable ways of getting polyunsaturated fatty acids into our diet hardly seems like progress.

THE PROBLEM WITH FAT

High fat diets may well have caused weight gain in some people, because fat is uniquely palatable and calorie dense. Trans fats probably had a significant effect on obesity (and heart disease), but this is now being addressed. Omega-6 oils do not cause obesity, despite what many might claim. And sadly, misunderstandings and years of poor advice regarding cholesterol badly damaged public trust in nutrition science, and left people not knowing what to believe.

There is a lot to take in here, and yet I have barely scratched the surface. Dietary fat is a deeply divisive subject, and prompts a lot of debate in the world of nutrition science. I doubt that there is a type of food that is the cause of so much contention and disagreement.

Apart from one, of course. Because the arguments about fat pale into insignificance when compared to the most divisive food of all. And that's carbs.

13
IS IT BECAUSE OF CARBS?

I think I can wipe out diabetes.

Robert Atkins

Carbs make you fat. Everyone knows that. It is the simple new dietary mantra, spouted by a million fitness gurus, self-proclaimed weight-loss experts and slimming celebrities. But given what we know about fat being uniquely palatable and calorie dense, can it really be true? And is it true of all carbs, or just some of them? Are there good carbs and bad carbs?

Certainly there are a number of passionate advocates for the 'it's carbs, stupid' theory of weight gain. Some even go as far as claiming that carbs are the root cause of the obesity epidemic. Advocates include athletes, doctors, dietitians and even a number of serious researchers. Here are a few quotes from some of the highest profile supporters of the carbohydrate hypothesis.

Any diet that succeeds does so because the dieter restricts fattening carbohydrates.

Gary Taubes, journalist

High fat/protein, zero carbohydrate diets have a dramatic impact on weight loss. This is because a diet with no carbohydrates in effect stops the production of insulin. If you eat nothing but fat and protein, for even a short period of time, despite the fact that your calorie intake may be high, you will see rapid weight loss.

Zoe Harcombe, diet book author

Carbohydrates . . . are not essential nutrients but powerful endorphin-activating drugs that are not controlled by the genetic hunger-satiety feedback system.

Professor Tim Noakes, diet book author

If carbohydrates are eaten throughout the day . . . in meals, snacks, and beverages, then insulin stays elevated in the bloodstream, and the fat remains in a state of constant lockdown. Fat accumulates to excess; it is stored, not burned.

Nina Teicholz, journalist

If all these people are correct, the debate is over. Carbs make us fat, and so reversing the obesity epidemic is simple. All we need to do is tax potatoes, bread, pasta and rice. Or, if that fails, perhaps ration or ban them from sale. We could then offer financial incentives for people to buy oils, fat, meat and fish. If we can rebalance people's diets in favour of high-fat consumption, we would see the pounds drop off. If the reason for obesity is really that simple, then so is the solution – let's get to the nearest KFC for a family bucket with no chips and stop playing around.

Of course, even if we really did succeed in making people cut carbs from their diet, there would still be a big fat problem to deal with. The only real sources of energy the human body can use are fat, protein and carbohydrate (technically alcohol counts as an energy source, although few would consider vodka an advisable way to take on calories). And so, dramatically limiting our consumption of carbs is likely to hugely increase our intake of protein and fat. As protein is often not the most palatable or delicious source of energy, a diet low in carbs is likely to be one that is high in fat. And with the risks of high-fat consumption discussed in the last chapter, that might not be ideal for people's health. This conundrum is the source of much of the debate, and has led to many of the arguments about fat that we have already discussed. Most advocates of a low carb diet also support high-fat consumption, and claim the evidence against it is weak. The UK cardiologist and diet book author Aseem Malhotra has long advocated cutting carbohydrates from our diet, and has gone as far as saying, 'Eat fat to get slim. Don't fear fat. Fat is your friend. It's now truly time to bring back the fat.' It is curious that the same people who insist that carbs are simply evil also manage to be convinced by the evidence that fat is simply good.

Setting the fat argument aside, what exactly is the evidence against carbohydrates?

THE ATKINS DIET

American cardiologist Robert Atkins became interested in low carb diets after reading a 1958 paper by Alfred Pennington entitled 'Weight Reduction'. Pennington's work seemed to suggest that if overweight patients were placed on a diet low in carbohydrates yet high in protein and fat, they would lose weight, even when they were allowed to eat as much as they wanted. In his groundbreaking 1972 book, *Dr Atkins' New Diet Revolution*, he claimed that high-fat diets had a metabolic advantage over other dieting strategies, and that burning fat uses up significantly more calories than burning carbs. The book, and many of the subsequent editions, was enticingly subtitled 'The High Calorie Way to Stay Thin Forever'.

Although low carb diets had existed before, it was Atkins's book that popularised it as a weight-loss strategy. All the other diets involved dreary, joyless, tiny meals, but with this one, you could wake up and eat a full English breakfast of buttery scrambled eggs, bacon, mushrooms, tomatoes and black pudding, leaving out only the toast. And the real miracle was that for many, the diet actually worked. People did seem to lose weight, and sometimes quite dramatically.

As the popularity of the diet grew and US rates of obesity began to rise, it raised the question: if cutting carbs makes you thin, could it be that carbohydrates are making everyone fat? Atkins certainly seemed to think so. As the size and scale of the increases in obesity became clear, he went as far as saying, 'How much obesity has to be created in a single decade for people to realise that diet has to be responsible for it?'

By the late 1990s, the Atkins diet had already grown from a popular book into a weight-loss empire. In the early 2000s, the diet's popularity was at its peak, adopted by millions of Americans struggling with their weight and desperate for answers. By 2004, a reported one in every eleven adults in the US was thought to be following a low carb diet similar to Atkins,[1] a shift so significant it had a serious effect on sales of pasta, bread and other refined carbohydrates.

The Atkins diet was helped by a number of benefits it has over more conventional weight-loss strategies. Firstly, it allows dieters to eat in an unrestricted fashion and still lose weight, suggesting that some sort of magical metabolic secret has been revealed. With only three macronutrients to play with, fat, protein and carbohydrates, there is only so much dietary manipulation that can be performed. So the idea that something so simple and achievable might be the answer is incredibly tempting to our simplistic brains. It gives us an illusion that we can take back control of our unruly bodies and life-hack our way to thinness, without any need for hunger and denial. We can be the thin person society tells us we should be, without pain or suffering.

The second advantage proved to be of enormous commercial value. Atkins, and other similar low carb protocols, is a diet that appeals to men. The low fat, calorie-restricted worlds of Rosemary Conley and Weight Watchers are very much female domains. They are all balance, health, slow losses and aerobic exercise in leotards. But with Atkins, here was a diet where you could eat real men's food. Steak, chicken, butter, bacon, eggs and cheese. As long as you followed a few simple rules, you could feast until you could eat no more, and still lose weight.

The third great advantage of Atkins is a little biological trick built into the diet that is both a fantastic marketing tool and a terrible curse. Although most of our body's excess energy is stored as fat, we also keep a relatively small amount as glycogen. Glycogen is made from long chains of glucose molecules, with some stored in our liver, and some in our muscles. We only have a limited amount of glycogen at any one time – it is not the most efficient way of storing energy, but it does give us a useful short-term supply of easily mobilised glucose.

Glycogen is converted into glucose when we aren't getting enough carbs from our food, so for anyone who significantly cuts carbohydrate from their diet, their glycogen stores are likely to be used up very quickly. Although all the glycogen stored by an average person probably only weighs about 500 grams, because of its complex structure, it holds onto a lot of water. And when glycogen is converted into glucose, that water is quickly released. As a result, one of the initial reactions to a carbohydrate-restricted diet is to lose a large amount of water. In the first week of carbohydrate restriction, dieters might eject up to 3 kilograms.

For many, this will be the most significant short-term weight loss they have ever experienced. To see the scales moving so far so quickly can be highly motivating. Most diets are full of long-term drudgery, with 1 or 2 kilograms being lost in a month. But the early results from Atkins are in a different league. Sadly, the water loss through glycogen depletion is a one-off, only occurring in the first week or so, and is obviously irrelevant when it comes to body fat. As long as carb intake remains low, the glycogen stores will not return, and the dramatic results will not be repeated. But if carbs are reintroduced, there is likely to be a significant short-term increase in weight as the stores are built back up and water is trapped, perhaps reinforcing the commonly held belief that carbohydrates make you fat.

For those who have struggled with calorie-restricted diets in the past, Atkins seems like it actually works. Well, sort of. The most popular and widely accepted view is that low carb diets see success in the same way that any restrictive diet does, resulting in exactly the same sort of temporary weight loss. Low carb dieting forces people to create rules around their eating habits, and omit many of their favourite food combinations. So instead of fish and chips, you just have fish, and eat fewer calories. Instead of pasta, vegetables and chicken, you have chicken and vegetables. Instead of a cheese sandwich, you have a cheese salad. When you order a burger, you throw away the fries and the bun. In some cases, people can't afford to overeat without cheap carbs to bulk things out.

The truth is, any strict rules about food will make you eat less and lose a bit of weight. If you only eat purple foods, only eat out of a teacup, only eat things with fewer than forty-six chromosomes, only eat standing on your head or only eat things you can pick up without your hands, you will lose weight in the short term. The same is true if you only eat low fat foods. Although there are some studies that seem to show low carb diets work better for long-term weight loss, there are plenty of others that show low fat diets are superior. And yet more showing we should follow vegan diets, or Mediterranean diets, or high-fibre diets. Everyone with a particular dietary philosophy to sell will be able to show you some evidence to support their claims. Trying to pick through them all is pointless and tiresome. They are all diets, and all suffer from the same disastrously low chances of long-term success.

I am also not about to start distinguishing between the many factions of low carb dieters that now exist, a rise that sent the once dominant Atkins Empire into administration in 2005. Paleo dieters claim to have discovered the secret to weight loss locked within our genes, but what they have actually discovered is Atkins with a blokey caveman fantasy. The many different commercial LCHF (low carb high fat) diets are just Atkins by another name. The Dukan diet is the same old low carb story, but with extra helpings of sliced ham. The ketogenic diet is a severe form of low carb diet where carbohydrates are brought down to incredibly low levels, attracting a special sort of dietary extremist. There are even organisations such as the UK's Public Health Collaboration, pushing the benefits of low carb dieting as a strategy for the nation's health, trying to pass off press releases as scientific research. But they can all be traced to the commercial success of the Atkins diet, and the many anecdotal stories of weight loss. Often there is an evangelical tone to the messages being spread, as if carbohydrate restriction has created some sort of religious conversion in these new gurus. All are convinced that theirs is the one true path, and others are fools not to follow.

I am certainly not interested in pulling apart every one of these very similar low carb diets as that would be akin to discussing the difference between fur and hair. But one thing worth investigating is the claim made by Atkins, paleo, LCHF, keto and the many low carb advocates that I quoted earlier. Within the wider low carb community, there is a commonly adopted belief that their diet has a 'metabolic advantage' over all the others – that there really is one true diet to rule them all.

Many believe that low carb diets do more than simply make people eat less food. Their hypothesis is that something specific about carbohydrate consumption drives fat into our adipocyte cells, making us gain weight. And beyond the world of silly fad diets and weight loss plans, there is a common suggestion that this unique property of carbohydrates is the main reason for the current 'epidemic' of obesity.

This is not just the opinion of a few dietary extremists. There are a small number of serious academics who strongly support these ideas. And their beliefs are not just based on a few compelling anecdotes, or the power of swimwear selfies. There is some science involved, and at times the arguments can seem very persuasive. It all starts with insulin.

BLOOD SUGAR

Insulin is an important hormone that has a number of different functions in the body. As we have already discussed, the three main sources of energy that we take on as food are carbohydrate, fat and protein. These reach our bloodstream as glucose, fatty acids and amino acids respectively, which are used as building blocks for our cells and as sources of energy. We have the ability to switch between fuels depending upon circumstance and availability, giving us a remarkable adaptability to different diets. The mechanisms that make this adaptability possible are very complex, but at the heart is the function of insulin.

Glucose is the favoured fuel of most of our cells, and so it is important that the levels are not allowed to drop too low. Like cholesterol, glucose is so important that our body can make it itself should dietary sources run low, in a process that occurs in the liver and is called gluconeogenesis. As with cholesterol, in our evolutionary history, this ability would have given us an important resilience in times of shortage. Carbohydrate cannot be found in any significant quantities in meat or fish, so almost all has to come from plants (although it is also likely that some populations of humans did get a large amount from honey). The capacity to make glucose was probably essential for our survival, giving us the ability to live on meat alone should circumstances require us to do so. For early inhabitants of Northern Europe, there would have been long periods of winter when plant food would have been in short supply. And for populations in the high Arctic, there would have been little dietary carbohydrate available at all.

One of the main roles of insulin is to regulate levels of the three fuels in our bloodstream. Like many things in the body, the preference is to keep everything as constant as possible, no mean feat when our food intake is so variable. If we eat a large amount of carbohydrate, as we digest it, lots of glucose will enter our bloodstream. When it does, our pancreas will start producing insulin, which will help to bring the glucose in our blood back down to a normal level. It does this in a number of ways. Firstly, it will stop our glycogen stores being released, and shut down any gluconeogenesis. Secondly, it will ensure that glucose is the only fuel being used by our cells, turning off any fat burning that might be going on.

This is at the heart of the carbohydrate-insulin theory of obesity.

Carbs cause insulin to rise, insulin inhibits the breakdown of fat into energy, and also increases the amount of fatty acids being produced (glucose is used as a building block when we make our own fatty acids, and so this is a good way of using up any excess). As insulin is produced in response to an abundance of food entering the body, this makes perfect sense. High levels of blood glucose mean that we probably want to shift to using glucose as fuel, and convert some to fat so we can store it efficiently. Insulin traps fat inside our fat cells, as it wants to lower the amount being released into our blood.

The more carbohydrate we eat, the more glucose enters our bloodstream, and the more insulin is released by our pancreas. When this happens, more fat gets made, less gets broken down, and more gets trapped in our fat cells. So, the hypothesis goes, carbs are a unique driver of obesity through the action of insulin. The journalist Gary Taubes popularised these ideas in his book *Good Calories, Bad Calories* and they have remained the stick with which people have been beating carbs ever since. For many, it is proof that obesity is in fact a simple problem with a simple solution.

In many ways, this is a compelling argument as there is no doubting that insulin does stimulate fat production and storage. But there are other factors we should take into account.

If you forgo dietary carbohydrates and instead eat a large amount of fat, it will result in fatty acids entering the bloodstream. Instead of glucose, your body will start burning fat as fuel. As well as an increase in the amount being used by cells, there will also be a decrease in the fatty acids being produced, an unsurprising reaction to getting plenty in your diet. But the reality is that simply running on a different type of fuel will not affect the amount of fat you have in storage. The body will happily run on either glucose or fatty acids, using as many calories as required and storing the rest as fat.

Even the most dedicated low carb advocates do not think that carbohydrates can magic extra calories into existence. So unless insulin drives us to eat more calories, or makes us burn less energy, it is impossible for it to cause us to gain weight. In response, the low carb world has suggested that insulin does somehow cause us to eat more, driving us into a state of so-called internal starvation – the idea is that when insulin is high, as we cannot access the fat we have in storage, we crave fatty acids and are driven to eat more food than we need. The hypothesis also suggests that this internally starved state causes a drop in resting metabolism in order to conserve energy, similar to the effects seen when leptin levels are low. This compelling idea has been popularised by a number of writers and diet book authors, and forms the scientific basis for the carbohydrate model of obesity. Many think it explains the age-old claims of Robert Atkins about a 'high calorie way to stay thin'. Cutting carbs makes you less hungry and crucially leads to you burning more calories. So is it true?

DOES INSULIN MAKE YOU HUNGRY?

Many are convinced. But sadly, the idea does not seem to stand up to scrutiny. Crucially, insulin does not seem to have the predicted effect on appetite. Numerous studies have shown that insulin actually decreases the amount that we eat, which you might expect from a hormone that is produced in response to an abundance of food. When the pancreas is stimulated to produce insulin, it simultaneously releases a powerful appetite-suppressing hormone called amylin.[2] [3] [4]

The internal starvation idea also suggests that weight gain will be associated with low levels of fatty acids, but again this appears not to be the case. In fact, obese people tend to have extremely high levels of fatty acids in their bloodstream, the exact opposite of the predictions.[5] It seems likely that gains in fat storage are the result of more fatty acids entering the adipocytes than leaving, not by them becoming trapped in fat cells by insulin. When levels of fatty acids in the blood are lowered using drug treatments, something that should mimic the 'internal starvation' state, there is no increase in either appetite or food consumption.[6]

It is also worth noting that insulin is not produced solely in response to carbohydrates. As I mentioned before, it is a hormone that controls and regulates the three different fuels our body uses. As well as responding to blood glucose, it also ensures that our body reacts when a lot of protein is consumed. When we eat and digest protein, amino acids enter our bloodstream and insulin is released. This stimulates us to start using amino acids, for instance in the building of muscles. And just as with the response to glucose, it decreases the release of fatty acids from adipocytes.

Often the spikes in insulin from eating protein are just as significant as those created by carbohydrates and last just as long.[7] [8] This insulin release is thought to be one of the reasons why high protein foods satisfy our appetites better than most others. But for some reason, those that demonise carbohydrates fail to account for why the release of insulin in response to protein is not as harmful as the one caused by carbs, perhaps because otherwise they would end up recommending a diet based entirely on fat. I guess you would lose weight, as a diet comprising solely of oil, butter and lard would be really hard to eat, but you would feel horrible.

The only other way high-insulin levels could make you fatter would be if they lowered your metabolic rate, but in experiments we find the opposite happens. High levels of insulin are associated with a higher metabolic rate,[9] and the ratio of fat to carbohydrate in people's diets has been shown to have no effect on base metabolism at all.[10] [11]

BUT WHAT ABOUT SPIKES?

Throughout much of the low carb literature, there is a great deal of talk about the need to avoid rapid spikes in insulin. For this reason, refined carbohydrates such as white rice, bread, noodles and pasta are thought to be particularly fattening. In some of the less extreme low carb protocols, dieters are told to stick to foods with a low glycaemic index (GI), such as wholegrains, pulses and lentils, as these are absorbed more slowly and so are less likely to cause insulin to spike, and to avoid high GI foods such as potatoes and refined carbohydrates. If the carbohydrate insulin hypothesis is correct, it might be thought that a high GI diet would cause people to gain more weight. But again, this appears not to be the case. When factors such as palatability, fibre content and calories are controlled for, the GI of a diet has no effect on weight loss, even when studied over an eighteen-month period.[12] So although a bit of extra fibre might be good for you, it has nothing to do with insulin spikes or weight loss.

There is an argument that high GI foods can be extremely palatable and calorie dense, and might drive people to eat too much, but the same is true of anything high in fat. And although I might be showing my bias as a chef, I do struggle with the argument that says we should be completely avoiding foods that taste good.

The main argument I can see against high GI foods is that focusing your diet on a small number of ingredients is a boring way to eat when so much else is available. Embracing the complex flavours and textures of pulses, wholegrains and root vegetables is something we should celebrate, rather than guilting people into eating them as sad pasta replacements. There is a good argument that when we make healthy eating seem like hard work, people don't keep it up for long.

For the majority of healthy people, the effect high GI foods have on insulin levels has no impact on weight gain. Insulin is unfairly demonised as a terrible hormone that we should do everything we can to avoid. It is actually a powerful and important part of our metabolic control system, something that helps us channel the nutrients in the food we eat towards the place they are needed the most.

But perhaps the most damning evidence against the carbohydrate insulin hypothesis is something we have already discussed. Almost all the genes so far discovered that are linked to obesity are expressed in the brain, particularly around the hypothalamus. Most are known to be related to leptin and other appetite-control pathways, suggesting that how fat we are is determined by our brain cells, not our fat cells. If insulin is the main driver of obesity, it would certainly be expected that some mutations in the genes that control its production or action would show a link to body fat. And if fat cells are the place where obesity originates, then surely many genes related to being fat would be expressed there. Apart from the incredibly rare mutation that stops people producing leptin, this is just not the case.[13][14]

Huge resources have been spent investigating the genetic basis of obesity and no relationship to insulin pathways has been found. This cannot be put down to research bias against low carb dieting, as genes are screened using statistical tools well before their actions are known. By now, if there was a link, you would expect at least some genes involved in the production of insulin in the pancreas, or in the way that it acts on fat cells, to have been found to have an association with weight gain. The fact that this is not the case is completely incompatible with the belief that insulin drives obesity.

THE CASE AGAINST CARBS

With so much evidence that carbohydrates and insulin do not increase appetite or lower energy expenditure, and no genetic basis to link them to obesity, why does the idea that carbs make us fat still persist?

One of the key factors that causes people to believe in the low carb hypothesis is timing. The correlation between the dramatic rise in obesity and the changes in the dietary guidelines is almost exact. The moment the US government told everyone to eat less fat, obesity started to climb. And when the UK and numerous countries around the world issued similar guidelines, the same thing happened. This is a commonly held view, represented by these quotes:

> The dietary guidelines changed in 1977 and in 1978 the obesity epidemic begins in the United States, and no one will take responsibility for that. And that's the question you have to ask . . . You change the guidelines and why won't you take responsibility for what happened? Why do you ignore it? And, why do you attack us for asking that question?
>
> Professor Tim Noakes

> This advice to avoid fat allowed the food industry to go hog-wild promoting low-fat, carb-heavy foods as 'light' or 'healthy', and that's been a disaster for public health.
>
> Professor Robert Lustig

Many others believe that the low fat guidelines were a catastrophic mistake, and the cause of significant sickness and disease. They are the root cause of obesity, and so the only sensible solution is to reverse them immediately.

There are many problems with this hypothesis, but perhaps the most significant is the number 30. As we have already mentioned, obesity did rise significantly around 1980, but only as a response to the steady and inexorable rise in average BMI throughout the twentieth century. As the epidemiologist Katherine Flegal noted, obesity is endemic and not epidemic. The only thing the increases in obesity represent are people crossing an arbitrary threshold of BMI, set at 30 because that was a nice round number. So the real answer to the question 'why did obesity start to increase in 1980?' is that someone at the WHO set the threshold at 30. If it had been set at 31, the rise in obesity would have occurred over a decade later and we would be blaming mobile phones, hummus and house music. If it had been 29, the increases would have happened in the late 1960s and we might point the finger at free love and LSD.

The only unique BMI change that occurred between 1980 and 1995 was a shifting of the bell curve to the right, meaning that a small percentage of the population became extremely obese. So did the guidelines cause that shift? Perhaps, but the carbohydrate insulin hypothesis does not provide a mechanism that might explain why a change in government advice would cause a very small percentage of the population to gain a lot of weight. Perhaps a more plausible suggestion is that the guidelines, the first in history that suggested we eat less of something, caused some highly susceptible people to become more anxious about their food choices, leading them to engage in restrictive eating behaviours, yo-yo dieting and a troubled relationship with food. In the next chapter, we shall discuss how this might affect some people in this way, and how it certainly has nothing to do with insulin.

DOES ANYONE LISTEN TO GUIDELINES?

In any case, did the change in guidelines actually have a significant effect on how people ate? And more generally, if we look at wide-scale carbohydrate consumption around the world, are increases associated with rises in BMI and obesity?

Again, the answer is a pretty resounding no. Although in the US carbohydrate consumption did increase after the 1970s guidelines, so did consumption of fat, and total calories. In the UK, fat consumption fell slightly, and so did carbohydrates. In Canada, fat consumption increased, both in total and as a proportion of all calories. Yet in all these countries, obesity rose significantly.

Although in the US, the most commonly cited example that supports the hypothesis, carbohydrate consumption did increase, longer term data shows a different picture. Throughout the twentieth century, as BMI has steadily risen, the consumption of carbohydrate has largely gone down, as people have become more affluent and eaten a more varied diet. Data from the US Department of Agriculture shows that between 1910 and 1970 carbohydrate intake steadily fell, before rising between 1970 and 2000. But even now, total calories from carbohydrate are nowhere near where they were in the early 1900s, when all but the very wealthy ate diets incredibly high in cheap carbo-hydrate rich foods such as bread and potatoes. It seems hard to understand how, if the carbohydrate insulin hypothesis is correct, obesity was almost unheard of in poor populations at the time.

The only macronutrient that rose consistently throughout the twentieth century was fat, with people including more into their diets as standards of living rose.[15] It seems likely that it was lack of poverty that caused people to gain weight, not the specific macronutrient breakdown of their diet.

Perhaps even more telling, when we look to data from countries such as China, the relationship between carbohydrates and weight gain is the exact opposite of what the insulin carbohydrate hypothesis would predict.[16] In 1962, an average Chinese person was getting around 80 per cent of their calories from carbohydrates, mostly refined white rice, something with a particularly high glycaemic index. By 2011, this had fallen significantly, down to well under 50 per cent, accompanied by significant rises in intake of fat and protein. And yet at the same time, the average BMI of the Chinese population increased significantly, with rates of obesity reaching unprecedented levels.

This is of course representative of the way that diets change around the world, as populations move from poverty to affluence. The world's poorest people get most of their calories from carbohydrates, because these are the cheapest sources of food. As they become more affluent, food becomes ever more available, particularly meats, oils and fats. People eat more of these previously unaffordable treats and their BMIs increase. What appears common to all these gains, across different countries, with different diets, is an increase in affluence.

THE HYPOTHESIS TESTED

One of the definitions of a hypothesis is that it can be tested by experiment. Clearly, however, one of the big problems in nutrition science is that people eat food in the real world, full of temptations, complexity and desires. It is well known that people lie about food consumption in self-reported surveys, especially if they have broken from a diet they are supposed to be sticking to, and this makes much data on weight loss prone to misinterpretation. If you really want to know the effects of a particular diet, you need to keep people in a laboratory and control exactly what they are eating all the time, and probably monitor their exercise levels and resting metabolism too. It would also help if you could accurately monitor changes in body composition, so you are not just seeing the effects of water or muscle loss. Sadly, though, this is expensive and difficult, and opportunities to do it are rare.

But they do happen. And when it is possible to perform these sorts of experiments, we should pay attention to the results. In 2015, Kevin Hall decided to test some of the predictions that the carbohydrate insulin model makes, performing a controlled feeding experiment in his laboratory. The diets of adult male volunteers were controlled across two residential stays, placing them onto either a six-day reduced fat diet, or a six-day reduced carbohydrate diet, and then measuring the effects. Both diets contained exactly the same number of calories, and physical activity was closely monitored and controlled.

If the carbohydrate insulin model is correct, clearly the group on the low carb diet should have seen a significant metabolic advantage, burning more calories and losing more weight. But these predictions were not played out. In fact, the higher carbohydrate version of the diet resulted in the loss of more body fat, something hard to reconcile with the idea of internal starvation.[17] Dr Hall told me: 'Under the carbohydrate insulin hypothesis, a reduced carb diet should result in more body fat loss and greater calorie expenditure since insulin secretion should be lower than when eating the reduced fat diet. But the results seemed to show the opposite. Of course, there was plenty of criticism that the experiment was not long enough and that the diet was not low enough in carbohydrate.'

Dr Hall had been discussing the experiment and the results with Gary Taubes, whose Nutrition Science Initiative (NuSI) was looking to fund studies on how the macronutrient content of diet affects weight gain. Taubes agreed that his NuSI would fund a further experiment in Dr Hall's laboratory, in order to address some of the criticisms of the previous work, and hopefully support Taubes' well-known views on carbohydrates.

In 2016, Kevin Hall's team performed the longer study, this time keeping groups of volunteers on the metabolic wards for eight weeks. For the first four weeks they consumed a high carbohydrate, high sugar diet, before switching to a very low carbohydrate ketogenic diet for the remainder of the time. Metabolic rates and body compositions were regularly measured throughout, using the lab's state of the art technology to get incredibly accurate results. If internal starvation was occurring, this study would surely find it.[18]

But even here, in a study funded by its most famous advocate, the results did not support the hypothesis. Dr Hall told me: 'We found no meaningful differences in calories burned. There was a small bump in the early stage of the ketogenic diet, but this quickly returned back to normal, and was far lower than the carbohydrate-insulin model predicted. Basically these studies counter the predictions and falsify the model.'

In this carefully controlled experiment, the extremely low carb ketogenic diet produced no detectable difference in body fat loss over four weeks, and only the slightest difference in energy expenditure. The amount this changed was barely detectable, even with the most sophisticated equipment available to nutrition science.

The arguments rumble on, and who knows, maybe people find carbs so delicious that they frequently eat beyond their appetite, in a way that they don't when on high fat diets. Or maybe after five weeks on a ketogenic diet, some as yet undiscovered adaptation kicks in, creating a huge metabolic advantage. But at some point, there has to be a limit. When a hypothesis makes testable predictions, and those predictions are not played out in controlled experiments, the only sensible conclusion is that the hypothesis is false.

In many ways it is a shame to see this idea falsified, because if things really were that simple, it would make life a lot easier. This is perhaps why so many are inclined to believe it is correct, even in the face of a large amount of contradictory evidence. We all like a simple story with a single hero and, ideally, a single villain. But a lot of simple explanations fall apart under close inspection. Accepting that things are complicated is the key to moving forward.

And so, although many people might benefit from a low carbohydrate diet, any suggestion that carb cutting is a solution to population-level obesity is deeply flawed. But increasingly, diet books are chock full of recipes with courgetti and cauliflower rice. We are encouraged to eat depressing low carb facsimiles of our favourite dishes, chucking yet more healthy and delicious ingredients on the enormous pile of evil foods.

Cut carbs if you want to, but do not suggest that others are inadequate for eating differently. And certainly don't pretend that you are hacking your bodily functions in some way, unlocking some metabolic secret based in nutrition science. Because after 100 years of manipulating macronutrients to try to make people lose weight, the only evidence we have is that it doesn't help.

I know that a lot of people will be screaming at this book right now. Because in trying to stick to a very specific discussion on carbohydrates and obesity, I have not mentioned two elephants in the room. One is a medical condition so intimately linked to carbohydrates, insulin and obesity that it would be ridiculous not to talk about it in this chapter. The other is the most demonised carbohydrate of all.

THE TROUBLE WITH FRUCTOSE

Most of the carbohydrate we consume is starch. Starch is composed of long chains of glucose molecules and our bodies use enzymes to break this down into glucose. This glucose is then used for energy and, if there is any spare, it will go to our liver to be converted into fat. So as long as we don't eat too much of it, we won't get fat.

But there are different types of carbohydrate that can enter our bloodstream, and because only glucose is used by our cells directly, these are processed in different ways. Perhaps the most significant of these is fructose, something that can form a large part of our diet if we eat a lot of sweet things. Refined sugar is made of a disaccharide* called sucrose, which breaks down into a glucose and a fructose molecule, which enter our bloodstream in equal quantities. These two sugars have the same amount of energy per gram as all carbohydrates, but behave in the body in a different way. And many think these differences cause sucrose, and particularly the fructose that we get from it, to be uniquely harmful.

Although the majority of glucose we consume will get used by our cells, all the fructose ends up in our liver for processing. Here it is converted into either glucose or fatty acids, depending upon what the body needs. Crucially, fructose in our bloodstream does not directly affect the production of insulin. Insulin can trigger leptin release to decrease our hunger. So if you give people large amounts of fructose, their leptin levels fall, they become hungry and the metabolism drops to conserve energy. However, this effect does not last. After four weeks of feeding with fructose, people end up with high levels of leptin, presumably because their fat storage has increased from the extra calories, but also because leptin resistance is starting to develop.

* Meaning it comprises two (di) simple sugars (saccharides).

And leptin resistance is thought to be one of the main causes of obesity.[19] So perhaps fructose really is the demon carb that everyone thinks it is.

Fructose consumption does appear to be associated with a number of problems, particularly in the liver where it is metabolised. Consumption of large quantities seems to result in the development of Non-Alcoholic Fatty Liver Disease (NAFLD), something that is strongly associated with metabolic syndrome. When the liver converts excess fructose into fat, it appears that some of it gets trapped there, forming dangerous fatty deposits. Fructose is also thought to contribute to the development of insulin resistance, and is strongly implicated as a cause of type 2 diabetes.[20]

These ideas have been brought to the attention of the public by the work and writings of US paediatric endocrinologist Professor Robert Lustig. He famously refers to fructose as the 'alcohol of the child', and claims that no amount of it is safe to eat. His 2013 book *Fat Chance* popularised the demonisation of sugar and inspired a thousand depressing low-sugar cookbooks, including his own.

Lustig strongly implicates sugar in the development of obesity. In a recent interview he explained how once fructose has caused fatty liver disease, it leads to further problems: 'Now you've got fatty liver disease. Your pancreas has to make extra insulin to make the liver do its job. Now you've got high insulin levels everywhere. You're gaining weight and your insulin is blocking the leptin in the level of the brain – making you hungrier. You've got a vicious cycle of consumption and disease.'[21]

With all the evidence pointing towards fructose being uniquely damaging, it does seem strange that there are not more recommendations for people to avoid it completely. After all, it metabolises using similar pathways to alcohol, with the potential to damage the liver in almost exactly the same way.

These problems are perhaps compounded when High Fructose Corn Syrup (HFCS) is used to sweeten foods and drinks. HFCS is a sweetener derived from corn starch comprised of both fructose and glucose in roughly equal quantities. It is commonly used in the US, where corn subsidies and taxes on imported sugar make it an extremely cost-effective option for drinks companies. The most common variant is HFCS 55, which has 55 per cent fructose to 45 per cent glucose, meaning that is has more fructose than conventional sugar. There are some variants with even higher fructose contents, but these are pretty niche, and are often diluted to make HFCS 55. In the UK, and most other countries, HFCS is barely used at all, mostly because sugar produced from domestic beet is a much cheaper option.

In 2004, the US researchers George Bray and Barry Popkin published a highly influential study showing a correlation between the consumption of HFCS and increases in obesity, suggesting that drinks sweetened with fructose played a direct causal role in its rise. This led many to conclude that fructose was the root cause of obesity, and corrupt nutrition authorities had caused everyone to get fat.

The conspiracy theory goes something like this. US corn subsidies led to overproduction. Desperate to find a use for this excess, food manufacturers developed the HFCS manufacturing process. In order to create a market for this new product, dietary guidelines were rigged in favour of low fat, high sugar foods. Americans developed a sweet tooth, and the huge amounts of fructose they were consuming made them all fat. The world followed suit, although everyone else became slightly less fat because we were all eating sugar instead of HFCS.

With both a plausible mechanism and epidemiological evidence, the case against fructose seems pretty damning. So why is it that sugar remains in our food supply? In the 2015 UK Scientific Advisory Committee on Nutrition (SACN) report on carbohydrates, strong recommendations for a reduction in consumption were made. But these recommendations were largely due to the known risks of dental problems, not some unique liver destroying properties. Despite reviewing all the available evidence, this committee comprising some of the UK's most eminent nutrition scientists found no association between sugar, BMI and body fatness. Even though there is very little love of sugar in the dietetics and nutrition communities, there are very few who agree with the sort of wholesale demonisation carried out by charismatic bestselling authors like Lustig and Gary Taubes. So why all the confusion?

Dr Tatiana Christides is a senior lecturer in human nutrition at the University of Greenwich. She is no friend of large food corporations, but she is one of many who get deeply frustrated by the popular narrative that sugar is the sole cause of obesity. When I spoke to her, she told me:

People tend to like simple explanations and struggle to accept that the answer is complex and requires high level changes. People like to focus on one nutrient to solve the obesity problem. It's sugar now, but it used to be fat. I feel that at times Lustig ignores the evidence that doesn't fit his theories, which I find shocking. It may be that you see increases in peripheral lipids after excess fructose intake, but we just don't eat fructose in the sort of significant quantities being studied, and we don't eat fructose alone. In some of these studies, people are getting 30 per cent of their energy from fructose. It's like . . .

She picked up a cup.

. . . if I push this cup with my finger, it will not break. I can push it forever and it will not break. But if I hit it with a hammer, it will definitely break. The amount of fructose being used in these studies is like a hammer. Just because they break the cup does not mean it will break in normal life. The big confounder in all these studies is excess energy intake. There is very limited evidence for fuel partitioning* in humans caused by fructose; we aren't rats and we metabolise carbohydrates in a different way, therefore I don't feel animal study results are always directly applicable to us.

Dr Nicola Guess is a researcher and associate professor at King's College London with expertise in the prevention of type 2 diabetes. Her research focuses on the nutrient-specific effects of the condition and she has also worked as a clinical dietitian treating diabetic patients. She told me: 'Excess sucrose, especially in liquid form, can lead to insulin resistance and increased visceral fat. But that is with excessive intake, with around 25 per cent of calories coming from sucrose. If you are drinking a litre of Coke a day, no shit that is going to cause a problem, but our guidelines have advised restriction of added sugars like sucrose far below this level for the past thirty years.'

The endocrinology professor Gareth Leng has a similar opinion, telling me that 'data from animal studies on fructose versus other

---

* Fuel partitioning refers to the way different fuel sources are used by our body at any one time – whether they are burned for energy or put into storage.

sugars don't provide compelling evidence for fructose-specific effects, except at extremely high levels – but there are differences in metabolism and especially in timescale of treatments.'

As Robert Lustig frequently implicates leptin in the mechanism by which fructose causes obesity, I thought it might be worth asking Professor Jeffrey Friedman, the molecular geneticist who discovered its function. He told me:

> With respect to whether some nutrients are more likely to cause obesity than others, years ago, fat was believed to be the culprit and now it is carbs. There has been a suggestion that leptin is involved. The truth is, no one knows whether some sources of calories are worse than others and persuasive experimental evidence is lacking. In actuality, it would require a very large study to answer this question and for a variety of reasons. The biologic system that regulates weight exerts itself over a long time, perhaps one to two years. Any study would by necessity need to be long term and would thus be expensive and limited by compliance. Absent direct evidence, if you have a strong view either way about carbs vs fats, in my view you are closer to the realm of religion than you are to science.

We know that too much sugar is still bad, and the same SACN report that I mentioned earlier clearly indicated that many of us eat way too much. Even if most of the evidence is related to tooth decay, it's an extremely serious issue, which can have wider health implications. But calling a food toxic and implying that no amount is safe is just not a good way to help people improve their diets or their health. It might sell books, but it pushes people to extremes, making them feel unnecessary guilt and shame.

TYPE 2 DIABETES

Even if we drop the obesity argument, there is still one stick with which to beat sugar. And that is type 2 diabetes. The 2015 SACN report did find a link between high sugar-sweetened drink consumption and the development of this condition, the seriousness of which we have already discussed.

There is much intuitive sense to the belief that type 2 diabetes is caused by carbohydrates. After all, it is defined as a resistance to insulin, and insulin is produced in response to carbohydrate intake. So if we eat loads of carbs, we will produce lots of insulin, and develop resistance to its effects. But carbs, sugar and fructose are not the only story in town. And the insulin response to carbohydrates seems to be a massive oversimplification. Nicola Guess explains:

> Starch per se does not cause insulin resistance. In fact, the evidence suggests quite the opposite. It is true that if people are given continuous infusions of insulin for seventy-two hours, that causes insulin resistance. But that is a very unphysiological study, because the level of insulin is flat. With carbohydrate intake, that does not happen. Pulsatile insulin secretion – a natural oscillatory pattern of insulin release which is lost in type 2 diabetes – is actually important for the maintenance of insulin sensitivity. Therefore, the rise and fall of insulin following meals might actually be helpful in terms of preventing insulin resistance. Starch intake in the absence of weight gain does not raise fasting insulin.

It is also important to remember that there is a difference between what might cause the condition to develop, and what diet strategies might be used in treatment. Just because a low carb diet might benefit a type 2 diabetic patient, it does not mean that a high carb diet caused their condition to develop. Nicola continues:

> I have no doubt that a low carb diet can be very effective in the management of type 2 diabetes. If you restrict carbs enough, you can reduce blood glucose quite markedly. What might be quite appealing for patients is that the reduction in glucose with sufficient carb restriction does not require weight loss. But it can also be very hard for people to stick to. It might be physiologically better, but it may not work for some people. There are other ways that people can manage type 2 diabetes that make more sense in their lives.

In reality, the causes of type 2 diabetes are complex, and, similar to obesity, these are likely to vary among individuals. Gareth Leng explains:

There's cause for interest in fructose certainly, and especially cause to think that sugar intake might be particularly relevant for diabetes. However, the main weakness of human studies – observational/epidemiology and intervention studies – is genetic heterogeneity. Factors that might be important for some individuals might not be for others. We just can't treat individuals as all alike, and that's a problem for those looking for public health solutions. I'm glad I'm not a clinician.

Similar to obesity, risk is increased by a number of known gene mutations, suggesting complex and multifactored causes. Many think the condition might be caused when our adipocytes reach their full capacity, causing excess fatty acids to be released into the bloodstream. High fatty acid concentrations can inhibit how much glucose is taken up into cells and used as energy, reducing the body's ability to respond to insulin. If for some reason our liver becomes less able to metabolise fat, this might also cause increased fatty acid circulation, and so insulin resistance. This might explain why, although there is a strong association between type 2 diabetes and obesity, some people are genetically susceptible to develop the condition at lower weights. It might be that they have lower fat storage capacity in their adipocytes, a liver that is less efficient at processing fatty acids, or perhaps that they are particularly susceptible to fructose in their diet.[22] A low carb diet might help some avoid the condition, but for others, the increased intake of fat might overwhelm their liver and make it more likely. Until we can really identify people's risks, generalised advice to follow an extreme diet is useless at best, and extremely harmful at worst.

There is an argument against sugar, and against fructose specifically, but all the relevant research is reflected within standard dietary guidelines. When we are driven to extremes, we run into problems, and with carbs, there is a particular problem. Because not all carbs are equal, and if you try to avoid them all, you might well be excluding yourself from the benefits.

THE FORGOTTEN CARB

Other than joy, perhaps the best argument against avoiding carbs is what you might be missing out on. For in fibre, there is a carb of unique benefit. Fibre describes a group of carbohydrates that is not digested by human metabolism, and that travels down our digestive track and into our colon. There, they either pass straight through or feed our microbiome, and so potentially contribute to many aspects of our health. Fibre also has the lowest calorie density of any food we eat, around two calories per gram, and is known to contribute to making people feel full after meals.

The 2015 SACN report found evidence for so many health benefits of fibre that it recommended increasing the guidelines for consumption, and also found that the majority of us do not eat enough. Although far too numerous to go into here, proven benefits include reducing the incidence of obesity, type 2 diabetes and colorectal cancer.[23]

Nicola Guess told me how carbohydrates digested by our gut bacteria might actually explain some of the benefits of low GI foods, without the usual insulin response story:

> There are likely to be extra-glycaemic effects of carbohydrate intake. Short chain fatty acids produced by gut fermentation of carbohydrates can have effects that help control blood glucose such as improving insulin sensitivity. There is a real limitation of considering carbohydrates only in relation to the degree in which they raise post-prandial* glucose. Some low GI carbs might be beneficial for reasons that have nothing to do with their direct effect on blood glucose and insulin.

By far the best sources of fibre are wholegrain carbohydrates, particularly pulses, beans, wheat, oats, potatoes and brown rice. Although you can and should get fibre from fruits, vegetables, seeds and other sources, the levels in them are relatively low. Fibre is a carbohydrate, after all.

THE BATTLE RAGES ON

The science of weight gain is complex, but the closer we look, the more it seems that the ratio of fat to carbohydrate in our diet is irrelevant when it comes to weight gain. Perhaps this is why, after years of

* i.e. after eating.

looking, no consistent pattern has ever been found in the data. As the argument continues, I see two sides, both convinced they are right. I am a chef, and although I care deeply about the results, I'm not in a position to perform my own systematic review of all the evidence and reach a firm conclusion. So all I can do is pick a side.

In one corner I see the low carb advocates, evangelical about a particular diet. They insist they are correct in the face of evidence. They assume that the expensive diets they eat can provide a solution for all. They advocate food exclusion, strict rules around eating, and even attack those who don't agree with them. Tim Noakes and Aseem Malhotra have both publicly abused dietitians about their weight, claiming that their advice should be ignored because of their appearance. Another low carb advocate once compared having a patient advocate on an obesity panel to putting Jimmy Saville in charge of looking after your children.

In the other corner, I see serious academics and clinicians, calling for balance, a proper review of the data, an acceptance that things might be complex. They ask for a realistic approach to dietary advice that will fit with people's individual needs and an acceptance that food is more than just nutrients. They present flaws in the carbohydrate insulin model, and ask the low carb camp to explain them away. They do not offer simple solutions, and rarely offer books or diet plans for sale. To me, one side sounds like scientists, the other like salespeople. I know whom I choose to believe.

The 2015 UK government Scientific Advisory Committee on Nutrition report on carbohydrates was a thorough and comprehensive review of all the available scientific evidence. It took a committee of the UK's leading nutrition experts several years to produce, and represents the most complete current review of the subject. It concluded that carbohydrates have no association with cardiovascular disease or type 2 diabetes. It also concluded that carbohydrate intake had neither a beneficial nor detrimental effect on the BMI or body fatness of adults or children. Despite our obsession, despite millions spent on research, despite endless books, articles and debates, no association has been found.

We are incredibly successful omnivores. We desire carbs because doing so conveyed an evolutionary advantage. We desire fatty foods and protein sources for the same reasons. We do not have to choose which is best. Whenever we do, be it with the idolisation of one nutrient or the demonisation of another, we only end up lurching between different unbalanced diets. It is not fat that has increased people's weight, neither is it carbs or sugar.

As every chef and product developer knows, foods made only from fat rarely taste delicious. Equally, anything made from carbohydrates or sugar alone will not appeal to a sophisticated palate. It is only when we combine these two things that food becomes uniquely tempting. Maybe in all the years desperately searching for the answer in one macronutrient or the other, we have completely missed the big picture. Could it be that particular combinations of fat and carbohydrate are the real drivers of obesity? Foods that are delicious, highly palatable and incredibly calorie dense. Foods so tempting they are almost addictive.

Well, I say addictive. Because when we talk about being addicted to food, what is it that we actually mean? Can we genuinely become addicted to eating?

14

IS IT BECAUSE WE
ARE ADDICTED?

*New discoveries in science prove that industrial, processed, sugar-,
fat-, and salt-laden food – food that is made in a plant, rather than
grown on a plant, as Michael Pollan would say – is biologically
addictive.*

Mark Hyman MD

*Quite simply, food is not addictive; drugs are addictive. And food
companies are putting drugs in our food.*
Kima Cargill, author of *The Psychology of Overeating*

The idea of food addiction is a flawed idea.
Hisham Ziauddin, senior clinical research associate at
Cambridge University's neuroscience department

When you are writing a book like this, it is easy to forget that most
people make decisions about food based on how much they enjoy
eating it. Even as someone who has spent most of his life trying to
make things taste good, when you write chapter after chapter explaining hormonal drivers, or breaking food down into calories and macronutrients, it is easy to forget about one of the primary drivers underlying our food choices: a desire to eat things that we like.

I have always thought that my life's work, cooking and developing
recipes, was a noble pursuit, or at least vaguely worthwhile. We all
have to eat every day, and if I can make that experience a little more

pleasurable for a few people, then I feel like I am adding slightly to the amount of joy in the world. We rarely focus enough on the little things, the tiny moments and interactions, even though these shape our lives. I have spent my life trying to improve some of those moments, by helping people enjoy the food that they eat. Even my writing is motivated by a desire to strip away the guilt that so many people have when it comes to food, in the hope that this will help them enjoy eating a little more.

But what if I have made a terrible mistake? Perhaps in spending my life trying to make food more enjoyable, I have contributed to an epidemic of obesity? Although I personally have no problem in resisting food temptations, could it be that certain highly susceptible people are being pulled into a cycle of addiction? Have we become fat because our modern food systems have made everything too delicious, and some of us cannot help but eat to excess?

This troubles me a great deal. As a chef, creating something 'irresistible' is pretty much the ultimate goal. That is what you aim for – to make food that tastes so good, people will crawl across broken glass for another spoonful. Never once have I thought, 'oh dear, this is way too nice – I had better make it less pleasant or people won't be able to resist'. Within certain cost and nutritional constraints, I have always aimed to make things taste as good as they can. I am fairly sure that every chef alive would say exactly the same.

But have the tools and techniques available to us in this modern industrial age become so sophisticated that we have created a world full of food addicts, all helplessly dependent upon our creations? Has the brilliance of food product developers made the world fat and sick? Could it be that my many development chef colleagues throughout the industry and I are so good at our jobs that we have caused an epidemic? And if so, should we be fired, or offered a raise?

HEDONIC EATING

Some of our drive to eat comes from pleasure, some from hunger. The two are closely interlinked, which is why food is far more delicious when you have gone without for a while, but 'hedonic hunger' can be considered separate from the more primal drive for sustenance. It's that feeling when you have just eaten your last bite of a meal, and you know you are completely full, your hunger fully satiated. And then a delicious looking dessert trolley appears, and you somehow find a bit more room. Pleasure can easily trump fullness, and often does.

Our hedonic hunger tends to be focused on highly energy dense foods, usually combinations of fat and refined carbohydrate. There are probably good evolutionary reasons why we might be driven to consume these things in the absence of hunger, because the likelihood is they would have been extremely scarce throughout most of our history. It would make sense to get them whenever we can, and to eat way beyond fullness whenever the opportunity arose.

Dean Burnett is a neuroscientist and writer with an interest in how our instincts control our behaviour (something explored brilliantly in his book *The Idiot Brain*). I asked him about why this hedonic hunger might cause us problems when it comes to eating behaviours.

> The brain is not a logical organ and can overrule a lot of the body's more fundamental reactions. It is best at learning when things are more visceral. Abstract information about eating healthily can be learned, but it will be overruled by pleasurable experiences. The brain will value short-term pleasure over long-term gain, and struggles to exert discipline over things that offer immediate reward. High-calorie foods activate reward pathways and can alleviate stress.

Different people will have different inclinations to hedonic hunger, something that is partly under genetic control (remember the children whose hunger preferences were defined from birth), and partly through learned associations. Food is about far more than just nourishment. It is about identity, culture, family, social bonds, friendships and love. It forms the basis of our most important celebrations. It gives us comfort, familiarity and joy. The hedonic aspect, and the importance of certain foods, will vary hugely among people, but generally the most pleasurable foods will have certain characteristics. The question is, can this intense pleasure lead to addiction? And if so, has it driven rises in obesity?

CAN YOU BE ADDICTED TO FOOD?

Popular food lore certainly says that we can. In surveys, 86 per cent of people believe that food addiction is a genuine condition, 72 per cent of people believe food addiction causes obesity and 40 per cent suggest that they have an addiction of their own.[1] [2] Many of us will openly admit to being addicted to chocolate, or unable to stop eating a particular brand of crisps once the packet is opened. Some will go as far as suggesting that certain foods contain addictive chemicals, and evil manufacturers are trapping us into compulsive consumption.

Many people claim that they cannot have certain foods in the house, as they will be unable to resist them if they are around. Others go to great lengths to avoid particular items, believing that one bite will never be enough. So tempting is the treat, so powerful the pleasure response, that any conscious control will fly out the window. Like an alcoholic who must avoid a single sip, they are helpless, in danger of spiralling out of control. To the untrained eye, these all look like pretty good arguments for food being addictive.

But these sorts of popular belief are not enough to classify food addiction as real. In fact, there is considerable academic debate about whether or not food is addictive, with much disagreement and evidence pointing either way. Although this sounds like it is a semantic argument of interest to a few academics, the definition actually has wide implications. If food addiction proves real, then it will be classified as a mental illness.

If obesity is thought to be driven by a genuine medical addiction, the way in which it needs to be approached becomes radically different. Treatments would change from diet advice, to rehabilitation and therapy. Freely available foodstuffs would become dangerous agents of harm. Chefs like me would be considered *Breaking Bad* style food pushers, locking people in with our irresistible wares. Legislation and control over these addictive substances would become the only option. The need to classify which foods are potentially addictive would be paramount, so that they can be legislated against in the same way as alcohol and tobacco. Parents allowing their children to have a Dairy Milk would be abusers; cheesecake would be sold in plain packaging, covertly offered from under the counter in back alleys. Friends would ask you to pick up Jaffa cakes from the duty-free whenever you travelled abroad.

I guess in one sense we are all addicted to food by definition, in the same way that we are addicted to breathing. But as the quotes at the beginning of this chapter reveal, for many, the addiction model is self-evidently a serious issue.

When our behaviour overrides our hormonal appetite control, a number of things can happen. Although leptin and many other hormones make eating progressively less pleasurable the more we consume, sometimes the joy that food brings us simply talks over those signals. As a result, we eat beyond the limits of our appetite. Our hormone levels try to shut down our hunger even more furiously. But for people with an extremely strong hedonic response to food, this is unlikely to reduce the drive significantly enough to stop.

There is an argument that constantly overriding these hormonal signals might lead us to a level of resistance to them. Most obese people have extremely high levels of leptin in their bloodstream, and this could be because they have grown resistant to it through regular overeating. When people constantly respond to hedonic hunger signals, it might lead to a vicious cycle where people are driven to consume more in order to get the same response, and get less pleasure from food as they do so. These ideas are quite speculative, and leptin resistance is known to have a strong genetic component. But there is some evidence that it can increase with overeating, leading many to conclude that people can develop a tolerance to food. And tolerance is one of the hallmarks of addiction.

But what exactly classifies something as being addictive? One of the major authorities in the world is the *Diagnostic and Statistical Manual of Mental Disorders* (DSM), which, despite some pressure to do so, did not classify food addiction as a recognised condition in its recent fifth edition. There are many reasons for this exclusion, including the potentially far-reaching consequences of defining obesity as being caused by a mental illness. There is also some debate as to whether food addiction should be classified as a substance disorder, meaning food would be put into the same category as alcohol and narcotics, or as a behavioural addiction in the same way as gambling. Again, this distinction would have profound implications for how the addiction might be treated, particularly at a societal level.

According to the *DSM*, in order to classify something as an addictive substance, it must fulfil at least three of seven criteria. These state that:

1. It must be possible to develop a tolerance to the substance.
2. Withdrawal must be experienced once someone stops taking it.
3. The substance must be frequently taken in amounts larger than intended.
4. Users should persistently try and fail to cut down.
5. Lots of time is spent obtaining, using or recovering from the substance.
6. Other activities are given up because of the substance.
7. Use continues despite persistent health problems caused or worsened by the substance.

For many people, food appears to meet three of these criteria easily, which is why some claim that it should be classified as addictive. But food is a more complex case than drugs or gambling, because it is literally impossible to give it up.

To take the criteria one at a time, there is certainly no convincing evidence for food withdrawal in humans, and the development of food tolerance is extremely contentious.[3] Although overeating is common, it does not have the same potential for catastrophic harm that excess drinking, drugs or gambling might. People often unsuccessfully try to cut down on food, but the ideal amount of food is not zero. Few people spend much time getting hold of food, or eat for extended periods in a way that negatively impacts their life. And although many health problems are linked to obesity, and it can certainly hinder taking part in some activities, the reasons why people become obese are many and varied. Food consumption and obesity should not be considered as the same thing.

Perhaps the reason why it is so tempting to assign food as an addictive substance is our tendency to link weight and eating behaviour, rather than the genetic susceptibility that underlies most of it. Food addiction is seen as something people wear on their bodies, their fat outwardly signalling their problems to the world. Yet even the most ardent supporters of food addiction would admit that it is not the only reason why someone might become fat.

Despite the public perception, countless bestselling books and regular media articles, there is still no consensus that food addiction is a clinical disorder. Nor is there a universal definition of food addiction in the literature. It is a great source of controversy, with the debate not stopping just because a clinical diagnosis cannot be made.

There is much evidence beyond the *DSM* criteria that food might be addictive, which has been enough to convince many that this is definitely the case.

CHEESECAKE, COCAINE AND ELECTRIFIED FLOORS

Some of the most compelling and often repeated evidence for food addiction comes from animal studies. As I have mentioned before, much of our drive to eat is based on systems common to all vertebrates, so evidence should not be discounted just because the experiments have only been conducted in rats. But in drawing any conclusions, we do need to be careful.

Food addiction can be induced in rats, but this involves very specific conditions, and only happens when certain foods are provided on an intermittent basis. This effect has been shown for sugar, high fat foods, and for various combinations of highly palatable items such as chocolate and cheesecake. Essentially it works for anything that is dense in calories, delicious and easy to eat. In these specific environments, rats can be driven to high levels of addiction. They have been observed to get the shakes when going through withdrawal, and willingly walk across electrified grids, receiving painful shocks in order to get their fix.

In the case of sugar addiction, there is a commonly cited claim that junkie rats prefer sugar to cocaine, which I can only assume was the result of an experiment drawn up on a university press department's night out. In reality it means little, other than the creation of a sensationalist headline or two. Many animals, including myself, will favour sweetness over cocaine, whether they are addicted to sugar or not. Most are driven by a desire for sustenance, rather than a need to become wide-eyed and annoyingly overconfident. It is also worth noting that saccharine produces exactly the same addictive effect, indicating that it is the desire for something sweet, rather than any physiological process, that motivates them.

The many food addiction experiments in animals have led to a wealth of compelling newspaper headlines and provide much supporting information to help sell numerous 'quit sugar' diet books. Here are a couple from the last few years, all supporting the commonly held belief that food, and particularly sugar, is a powerfully addictive drug:

> In animals, it [sugar] is actually more addictive than even cocaine, so sugar is pretty much probably the most consumed addictive substance around the world and it is wreaking havoc on our health
>
> James DiNicolantonio in the *Guardian*

> Studies suggest sugar is eight times more addictive than class A drugs . . . The more sugar you eat, the more likely it has taken hold of your addictive pathways and is driving you to eat – and drink – far too much.
>
> Karen Thomson, *Sugar Free: 8 Weeks to Freedom from Sugar and Carb Addiction*

James and Karen are almost certainly wrong. Animal studies can show us a great deal, but the behaviour of addicted rodents is very specific, and only occurs under certain conditions. Without the intermittent access and strict control of the foods available, rats do not become addicted and this is totally unrepresentative of anything encountered in the real world. We are frequently told that one of the main problems with our modern food environment is our constant access to calorie dense foods, the opposite of what is required to make food junkies out of rats. More importantly, the behaviours in rats have never been observed in humans. We do not develop tolerance, nor show signs of withdrawal. And if sugar addiction was substance specific, presumably the most severe addicts would end up eating the refined product straight out of the bag, which I am not aware of, even anecdotally.

THE ADDICT'S BRAIN

There is, however, supporting evidence for food as an addictive substance from investigations into brain activity. Certain foods have been shown to activate the same reward centres as addictive drugs, and are involved in similar dopamine* pathways. But it seems that this

---

* Dopamine is a neurotransmitter (brain-signalling chemical) involved in driving a feeling of reward.

too is a vast oversimplification. Dr Charlotte Hardman studies the psychological determinants of appetite and eating behaviour at Liverpool University and has done a great deal of work on food addiction. She told me:

> People say sugar releases dopamine so it must be addictive, or certain foods target the same brain areas as drugs, so they must be addictive. But this is simply a misunderstanding of how the brain works. Lots of things activate the reward centres. They tend to respond to anything nice, motivating or relevant at that moment in time. Music that we like, or pictures of attractive people. They were even shown to respond to pictures of George W. Bush by his supporters around election time, but I don't think anyone would consider George W. Bush addictive.

> Drug effects are generally prolonged and stronger. The hallmark of addiction is an inability to exert control over your response. The frontal part of our brain controls or dampens down our response to reward. It's a bit like the rider taming the horse. Overeating will happen when there is a really strong motivation, and a lack of control response. Our work indicates that there is no specific addictive substance in food, but that some people have a strong motivational response to really specific items that have particular psychological significance. It might be a chocolate bar, a KFC or a burger.

When I asked if that makes eating these foods a behavioural addiction, she replied, 'You can't say it is like gambling. There are similar underpinning processes to gambling, but eating behaviour is completely different, so really needs different criteria.' The idea that certain delicious foods might light up our pleasure centres should not really be surprising at all. These highly motivating reward pathways have been designed to keep us alive. We are programmed to respond to sweet, fatty, energy-dense foods in this way because of our need for sustenance. Which is why drugs that tap into these pathways are so compelling.

Also troubling for supporters of food addiction as a cause of obesity is the way that pleasure centres light up in response to food in obese people. It might be expected that those prone to overeating would show a consistent response, but this seems to be far from the case. A

2012 review of food-addiction literature, conducted by a group of Cambridge University researchers, concluded:

> The vast majority of overweight individuals have not shown a convincing behavioural or neurobiological profile that resembles addiction. Indeed, the enormous inconsistency emerging from a review of the neuroimaging literature tells us that in this highly heterogeneous disorder, the application of a single model is likely to be more of a hindrance than a help to future research.[4]

And this, it seems, is the main problem with the concept of food addiction, and indeed any model for 'the cause' of obesity. It is really complex. Just because someone looks fat does not mean you know exactly how they arrived at that point. There may be a level of addiction to certain foods involved, but those foods are probably specific to the individual, defined by memories and preferences. There may be no addiction at all, with weight gain being caused by other factors. Equally, food addiction might be affecting many people who are not fat, causing them to overeat certain foods, but compensate in other ways.

To really understand if an addiction model is helpful, we need to be able to identify it, rather than making sweeping statements that fat people must be addicted, because, you know, just look at them. We need to consider if our desire to think of fat people as hopeless food addicts is not just playing into our prejudices of them as weak-willed and lacking control. Are we just framing our disgust in a more benevolent way, so we can feel sorry for them and pack them off for treatment?

In fact, binge eating disorder (BED), where people regularly eat to excess, often with associated feelings of guilt and distress, does satisfy the *DSM* criteria far better than other models of food addiction. And even though it is rare in the population, among people with the most severe cases of obesity it is surprisingly common. Ulrike Schmidt is a psychiatrist and professor of eating disorders at King's College London. She explained:

> There is certainly a subgroup of people with obesity who have Binge Eating Disorder. A middle-aged man might become obese if he is in a sedentary job, eating and drinking a bit too much. But this is very different to people with Binge Eating Disorder and obesity. These will be people dieting, then losing some

weight, followed by out-of-control eating episodes. About 30 per cent of treatment seeking obese people have Binge Eating Disorder. In the bariatric* population, usually the most severe cases, that rises to between 50 and 60 per cent. It is not as simple in humans as it is in rats, but it has elements of being addiction-like. Maybe it is because people are replicating the intermittent access to food that produces addictive behaviours in animal experiments. Because they want to rectify their weight by dieting they avoid calorie dense foods. Then, when they do get access, they binge.

If we agree that the biggest problem when it comes to obesity is the rise in severe cases at the far right-hand side of the bell curve, then it is possible that over half of those people are suffering from BED. It is also possible that this behaviour is being driven by restrictive dieting. Although our modern food systems allow us constant free access to calorie dense food, could it be that a culture of dieting is producing the very environment of intermittent availability that might cause addiction?

Even if this is the case, it would be foolhardy to think that ending the culture of dieting or a restructuring of the food environment would solve all the problems being experienced by people with binge eating disorders. A 2001 study found that 83 per cent of BED sufferers had a history of childhood abuse, 30 per cent had experienced sexual violence and 69 per cent had suffered from emotional neglect.[5] Binge eating disorder should not be thought of as a substance addiction, but evidence of a deeper malaise. When people are suffering, they need to be shown compassion. Instead they are abused and marginalised for being fat, told they are not worthy of a place in society. Newspapers claim they are drains on society, that they will never find love, that they are ugly and weak.

If there is one thing that makes dealing with trauma harder, it is having to deal with it alone. It is feeling that you lack worth and believing that society has no place for you. This is what we do to obese people, particularly the most severe cases. Instead of the compassion and inclusion that they need to help them through trauma, we push them away and compound their suffering. Many of the people responsible for doing this claim that they are just showing some tough love. Perhaps they do not even know how cruel they are.

---

* Peope who have had weight loss surgery. See Chapter 17 for more on this.

WHAT'S THE POINT OF 'FOOD ADDICTION'?

With so much uncertainty, the big question is, would the classification of food as an addictive substance, or eating as an addictive
behaviour, be in any way useful? The answer, once again, is complex.
Some studies have shown that framing obesity as a result of food
addiction might reduce stigma.[6] After all, the idea of food addiction
might go some way to showing that obesity is not a matter of personal
choice. But many feel that the dramatic change in approach needed
to tackle a newly created addiction crisis might create more problems
than it solves.

Charlotte Hardman has looked at how knowledge of food addiction
might alter people's behaviour.

> We did a study where people did a computerised task and filled
> in a questionnaire and were given false feedback as to where
> they sat on an addiction scale. Those that were told they scored
> high limited their exposure to certain foods, at least in the short
> term, but it might not be something that causes sustainable
> change. I do feel it might help people appreciate the complexity
> of obesity, but what we really need is an idea about how it might
> affect public perception.

There is no doubt that food, and our relationship with it, is different to
any other substance that we already consider addictive. Similarly,
eating is different to any other addictive behaviour. You can engage in
all the restriction you want, but eventually you will need to eat. If you
have denied yourself food, you will be more likely to lose control when
food becomes available.

If we reject the idea that there are specific addictive substances
contained within our food, as most of the sensible literature does, what
are we left with? An intense desire for certain delicious things that can
override our control mechanisms. Although this is more common for
energy-dense foods, it is largely governed by our feelings and associations. Some may lack the ability to override these desires, and so are
driven to compulsively eat, even though they know it is doing them
harm. When this happens, it is easy to see food as addictive, because
from the outside it seems like persistent and irrational behaviour. But
perhaps to the person eating, they are just enjoying the visceral

pleasure of food, without thought of future consequence. Perhaps they are more carefree than they are addicted.

Just what is it that creates this desire for certain foods? It is not a world that I recognise. And what is it that limits some people's ability to control themselves? Dean Burnett shared his thoughts on what might create such seemingly irresistible drives: 'Our brain has a tendency to indulge in forbidden thoughts and forbidden thinking. It likes to think about things that are not widely thought and accepted, and these thoughts become more and more powerful the more you try and suppress them. It is one of the paradoxical ways in which the brain works. Like the way you can't force yourself to relax.'

In our desire to be healthy, we have labelled chocolate, cakes, fried foods and other such indulgent treats as forbidden. With the best will in the world, calorie-dense manufactured products are not the most delicious of foods, yet they are the ones that drive people to overconsume. Personally, I enjoy a Twix as much as the next person. But I am capable of resisting its lures, even when it is placed right in front of me. If I am at a shop counter and someone offers me an unwanted discount, I will only say yes if I am hungry enough. If I have a Twix in the cupboard at home, I am not consumed by a desire to eat it the moment it catches my eye.

Perhaps this is because I have never dieted and so never felt compelled to resist. I do not think I have ever banned chocolate from my home, and I have certainly never structured my life in order to avoid it. Of course, it may be that I experience naturally low levels of hedonic eating. But given that I have dedicated my life and career to a love of food, I find this hard to believe. Perhaps I have exceptional willpower. But in many other areas of my life, I am often weak-willed and poorly disciplined.

Although my personal experience is no substitute for evidence, I do believe that in refusing to create a moral framework around these sorts of foods, I have unwittingly made my hedonic desire for them easier to control. And when we look at 'Intuitive Eating' in Chapter 18, we shall see there is a good deal of evidence to support this.

For anyone who believes a Twix to be sinful, and persistently forbids themselves access, perhaps this combination of caramel, chocolate and biscuit becomes impossible to resist. In creating a framework of guilt around so many foods, have we give them power beyond their sensory temptations? Studies of hedonic eating show that dieting and restriction create a state of psychological deprivation as powerful and motivating as any emotional starvation. For many, this will be a hunger that only certain forbidden fruits can feed.

We talk of people being unable to resist the temptations of the modern food environment, but this has more to do with guilt than the brilliance of development chefs (we're good, but not that good). High-calorie manufactured foods can be harmlessly enjoyed. It is our culture that makes them dangerous. In places where Twix and Dairy Milk do not occupy the role of guilty pleasures, they fail to dominate the supermarket shelves. The demand for these products comes from within us, and the only way this will ever change is if we alter our relationship with food. The alternative, where we fix the supply without addressing the demand, paints a cruel dystopia. Parents having their children taken from them for feeding them Haribo. Black markets, pushing the supply of Battenberg underground. Illegal streaming services offering old episodes of *The Great British Bake Off*.

However much you are dedicated to fighting obesity, you need to ask yourself what sort of world you want to live in. Before you insist that the food system needs to change, think hard about what the end game would look like. For many, it would not look like progress.

In a famous experiment, the legendary food psychologist Paul Rozin showed different groups of people around the world pictures of a delicious looking chocolate cake and gauged their reactions. In the US, the first thought that came to people's minds on seeing the image was guilt. In France, which at the time had far lower rates of obesity, the first word was celebration. Ironically, far from saving us, marking indulgent foods as 'bad' only seems to drive us to want them more.

15

IS IT BECAUSE OF OUR ENVIRONMENT?

The prevalence of obesity in the population generally varies by a number of factors, such as sex, age, race-ethnic group, educational levels and smoking status. The increases in obesity, however, occurred in all these groups fairly uniformly, suggesting that the factors that cause variation in obesity prevalence in the population are not necessarily the same as the factors causing the increases in prevalence.[1]

Katherine Flegal, epidemiologist and senior scientist
at the CDC National Center for Health Statistics

The only effective approach is for governments to implement radical policy change, to regulate food consumption and control the food industry in a similar way to that of the tobacco industry, by banning the advertising of selected produce, taxing certain foods, and rationing the purchase of others.[2]

Professor Rachel Davey, director of the
Health Research Institute, University of Canberra

While researching for this book, in order to get as wide and balanced a view as possible, I spoke to a large number of people from many different spheres. Perhaps one of the big problems with obesity, and the reason why it can be so complex and divisive, is because there are so many different areas of expertise that might be employed to explain it. I have spoken to geneticists, molecular biologists, economists, biochemists, neuroscientists, addiction researchers, psychologists, microbiologists, psychiatrists, philosophers, dietitians, nutritionists, sociologists, eating disorder specialists, medical doctors, endocrinologists, geographers and sensory scientists. I have spoken to those who dedicate their lives to helping obese people and to others who live with their own larger bodies every day. But even after all this, I feel I have barely scratched the surface. I suspect that there are many others with potentially valuable expertise and insights, perhaps from fields that I didn't even consider.

Everyone I spoke to had their own specialism and knowledge, and each gave me something unique and interesting to think about. Many of them are quoted in the text, but others simply helped me increase my understanding. Although I had differing lines of enquiry for everyone, I ended each interview with a couple of common questions. I asked everyone what they thought the main reasons were that obesity continues to rise and what might be done to bring it down.

One day, I may compile the answers to these questions more formally, because they are fascinating. But what stood out was that among most, but not all, of the people I spoke to, there was a common theme. Almost everyone agreed that certain aspects of our modern environment cause people to become obese. The world has changed, and it encourages us all to eat more and move around less. An 'obesogenic environment' is the main factor causing global shifts in body weight.

The availability and marketing of high-calorie foods is blamed by many as a key driver of weight gain, although others point the finger at our more sedentary lifestyles. The environment seems to have become uniquely obesogenic in recent years, and it does not seem to be improving. Unless we can somehow manipulate the world in order to change things, it is only going to get worse.

Obviously I agree with these insights. Interactions between our genes and environment define everything about us and that includes how much energy we store as fat. But sometimes a leap is too readily made to a couple of obvious factors that might be causing weight gain. Evidence that we are eating more than we used to is inconsistent and seems to vary among countries. And as we know from the exercise chapter, the idea that increasingly sedentary lives have made us fat seems troublingly simplistic at times.

Also tenuous is any idea that our diets have declined in quality. There have been very few major shifts in macronutrient intakes in the last 100 years, and none of these show consistent correlations with weight gain. What happened to diets in America is radically different to what occurred in China, Canada, Australia, South Korea and the UK, yet all these countries have seen significant increases in body weight. There is also plenty of evidence that our diets have been improving in recent years. Increasingly efficient food supply chains mean that we have access to more fruit and vegetables than ever before, and, in the UK at least, consumption has increased.[3]

A lack of correlation between diet quality and obesity is perhaps surprising, but really shouldn't be. Studies of factors affecting the risk of becoming obese show that, other than a small effect from eating more fibre, the nutrient breakdown of our diet is pretty much insignificant.[4] Increasing fruit and vegetable consumption, although it doubtlessly has many other health benefits, has not reliably been shown to have any influence on weight.[5] Food quality seems to be far less important than factors such as maternal smoking, genetics or ethnicity.[6]

But trumping all of these, and by a considerable distance, the greatest risk factor of all is living in America. It is estimated that when compared against other high-income countries, spending your life in the USA increases your risk of becoming obese by over two and a half times. America is closely followed by the UK and Australia in this regard, or in fact any country where US cultural values are closely adopted. It seems that culture drives weight gain to a far greater extent than diet, which is both curious and unsettling. It might be assumed that it is simply the economic advantages of this Americanisation that drive weight gain, but it seems that socio-political changes have far more influence. Changes in culture and society seem to have a much greater effect on people's body weight than affluence and free access to food.[7] Maybe we need to look deeper into what American culture means, and how living in a Westernised society affects people's lives.

As the quote from the epidemiologist Katherine Flegal explains at the beginning of this chapter, in the countries where it has occurred, increases in obesity have happened uniformly across people of different ages, sexes, ethnicities, and education levels. It has occurred within groups occupying vastly different environments, and eating massively diverse diets. There have been some recent changes across socio-economic gradients, but whatever environmental factors are driving people's weight gain, they are happening widely and persistently.

The environmental reasons for population-level weight gains are far more mysterious than most people admit. We blame food because it is an easy target. When that argument breaks down, we blame physical barriers to exercise, although given the absence of data for how activity levels have changed, most people would accept that this argument is fragile. The truth is, no one can really say for sure.

CHANGING THE ENVIRONMENT

An increasing number of people, in fact I would go as far as to say all reasonable and informed people, reject the idea that obesity is caused by a lack of individual responsibility. You have to be a special sort of idiot to imagine that people are fatter because they are less responsible these days, although as we have seen, expressing this view doesn't put you out of the running for UK Prime Minister.

The prevailing liberal view, and the one held by the majority of academics, clinicians and experts I have spoken to, is that increases in body weight have been caused by a combination of our built environment and our food system. The world we live in does not encourage exercise and activity, and at the same time it drives us to consume ever more calories. Modern economies require growth, and the only way for food companies to grow in a constantly satiated world is to encourage people to eat more than they need. Big business forces more and more calories down our throats, regardless of the cost to our health. Food marketing has made us fat.

This is known as the 'push hypothesis' and has been investigated by Dr Kevin Hall as part of his studies into how changes in calorie consumption affect population weight. It seems that people do consume more calories as they become available, and the increases are enough to explain the population-level changes in body weight, even when metabolic resistance is taken into account.[8] At least in the US, the population-level increases in stored body fat can be accounted for by extra calories entering the food supply. When excess food exists, food companies will find a way to sell it to us, whether we need it or not.

Manufacturers and retailers are forever devising increasingly cheap, calorie dense and tempting foods (sorry about that), and use cunning advertising and promotion strategies to suck us in to vicious spirals of overconsumption. With their increasingly sophisticated consumer marketing techniques, they have us in their thrall. Or at least they have fat people in their thrall. Presumably people like me are somehow immune to their charms. Which, when you say it out loud, does seem slightly problematic.

We shall return to this later. Firstly, it is worth thinking about where this argument leaves us. If the environment is to blame, the solution presumably lies there too. This idea dominates public health discourse. We need to change and control the food that people eat. The quote in the introduction from Professor Davey might be on the extreme end of things, but it is not far from the established view. The UK government's Foresight report into obesity complained how the public are bombarded by adverts and promotions for fatty, salty, sugary foods. It suggested that the problems being faced were huge, complex and hard to address, concluding that 'a substantial degree of intervention is required to affect an impact on the rising trend in obesity. A systemic or paradigm shift is needed to disrupt the cycle of accumulation of fat and to restore balance.'

Many countries are planning a tax on sugar-sweetened drinks, and as we have discussed, there is some evidence that these products can cause us to take on excess calories that are not compensated for. I have ideological reasons for being uncomfortable with taxation, partly because it is a regressive tax likely to disproportionately affect people on low incomes. But there is certainly evidence of a need to cut down on sugar, if only because of the associated dental problems, so sugary drinks be damned. I am a chef, and my focus, as always, is on food.

Still, if we are serious about changing the environment with a 'substantial degree of intervention', a few pence on a bottle of Coke is not going to cut it. We need some big changes and tough decisions. So, what's next?

How about chocolate? After all, it is uniquely palatable. Its melting point, just slightly lower than human body temperature, means that the sensation of eating it brings much joy. It is a near ideal calorie delivery system, comprised largely of fat and refined sugar. Its complex flavour means that it doesn't get boring or sickly in the way that buttercream or icing would. It is associated with love, celebration, comfort and reward. Some turn to it in times of stress, upset and low mood. It is the food that most fulfils the role of guilty pleasure in many people's lives. It is also widely available in a great number of brightly coloured and tempting options, often specifically designed to appeal to children.

If any single food is driving obesity, I would point the finger at chocolate. It is as close to being an addictive drug as any food there is. Give anyone, of any age, a choice between sugar and chocolate, and 99 per cent of the time, chocolate is what they will choose.

So what should we do? Placing controls on advertising chocolate to children would be a good start. Make sure that chocolate is never promoted during children's television programmes. Ensure that chocolate companies are not allowed to sponsor children's events, and that chocolate is not sold at their eye level in supermarkets in order to activate the famous pester power of leg-pulling toddlers. I think that most people would be comfortable with these measures, and, through a combination of legislation and voluntary implementation, this is not far off where we are in the UK. Unfortunately, it does not seem to have achieved a great deal. Although time will tell, ask a class full of primary school children to name their favourite food, and most will tell you how much they love chocolate.

What else might we do? Ban all advertising? After all, children watch many programmes that adults watch, especially sports. So maybe we should stop chocolate being promoted anywhere. We are not just worried about childhood obesity – we want to tackle it in adults too. But we would still see chocolate every time we went into the supermarket, sitting there in its bright packaging, shouting 'buy me'. We would all still know how good it tastes and how much pleasure it can bring. Many food brands barely advertise at all these days, and still sell millions of units. We might become less brand loyal, but I do not imagine we would stop loving chocolate.

Perhaps the most effective way to reduce consumption is to make chocolate more expensive. The problem with this is that the price of chocolate is extremely inelastic, meaning that when it increases moderately, sales are largely unaffected. So any small change in retail price caused by a 10 or 20 per cent tax might not have much effect on the amount being consumed. And even if it did, it would probably have virtually no impact on those who love chocolate the most, and who are at the greatest risk of harm.

What if we double the price? Are we still comfortable? Or have we just made chocolate unaffordable for people on low incomes? How long will we go on increasing the price? Until people stop being obese? Until half as much chocolate is being eaten? Until only the top 10 per cent of earners can afford to eat chocolate? Until the companies making it are run out of business and thousands are left unemployed?

If not taxation, how about reformulation? We could take calories out of our favourite treats, and make them better for us. Again, this is a laudable aim, and one that people tend to be more comfortable with. But even here, there are problems and limitations. If you take sugar out of chocolate and replace it with a low-calorie sweetener, the sweetener is generally used in far smaller quantities, and so the lost weight has to be replaced somehow. If fat is used to replace it, the chocolate bar will actually have more calories. If protein or another form of carbohydrate is used, there will be no change in calories (protein and carbohydrate have the same number of calories per gram). You could add some sort of dietary fibre as this has a lower energy density, but you'd quickly end up with a food product that has a strong laxative effect.

So if you want to reduce calories in chocolate, the usual approach involves dropping the portion size. This results in consternation and upset among the buying public, and often negative press as newspapers claim 'shrinkflation'. True, you can always reduce the price to avoid claims of ripping people off, but the reality of making a food product smaller is that much of the real cost – packaging, logistics, energy, labour, factory over-heads – remains constant. So a 10 per cent reduction in ingredients costs is unlikely to result in a 10 per cent reduction in price, or even a 5 per cent reduction. It might be suggested that manufacturers take the hit and put up with reduced profits. There is a good argument for this, but with markets as they are, eventually this would make businesses unsustainable.

And what would this reformulation or portion size reduction be likely to achieve? At best, maybe a 10 to 15 per cent reduction in energy content. For a 200 calorie chocolate bar, that represents a maximum of around 30 calories. Even if someone eats one every day, this is still a fairly insignificant amount of energy. Our bodies can, and probably will, adjust, and stubbornly remain exactly the same weight. You really need to stop people eating them completely. Which leads to regulation. Rationing. Banning chocolate from schools and homes. And when all this is done, I imagine the dark forces of 'Big Pizza' might have cause for celebration.

There is of course a point where everyone becomes uncomfortable with this sort of regulation. Well, almost everyone. In the paper that I quote from at the beginning of this chapter, Professor Rachel Davey sets out the case for such extreme measures. She says:

> To tackle the fundamental problem of positive energy balance, there needs to be radical changes in our society and the environment in which we live that enforce and empower people to change their eating behaviour . . . Legal actions against the tobacco industry have helped to break down their hitherto impenetrable wall of defence. There is no reason why this could not be so for the food industry if there is political will and financial commitment. Although many might recoil at the idea of food taxation and control, the tobacco precedents show that public opposition can, in time, be successfully overcome.

CIGARETTES, THAT FAMOUS NUTRITIONAL CATEGORY

The success of smoking interventions is the argument used to counter any queasiness people might have with the control of food. In response to initial calls for smoking regulation, the tobacco industry produced some very similar arguments to the ones I present here. But food is crucially different in many ways, making these sorts of comparisons salacious and unhelpful.

Firstly, no one has ever been harmed by passive eating. Or indeed by someone else becoming obese. Whereas the harm caused by passive smoking meant that the public's health was at risk from someone else's actions, when it comes to food, no harm is done to others through your eating habits (unless you are doing some really weird shit).

Crucially, with food, there is potentially huge harm in reducing the amount that people are allowed to consume. Different groups, particularly the elderly, have difficulty absorbing enough nutrients from their diet as it is, and so are placed in great danger from any forced restriction. The calorie dense items that cause weight problems for many can be a godsend for others. We all have different needs, determined by our genetics, life stage, general health and physiology. If we decide to regulate the nation's diet, how will we account for these differences? Who will decide what people can and can't eat? For people prone to eating disorders, behaviours are likely to be compounded by any restriction and rationing. Remember the rats that became addicted to calorie dense foods only when intermittent access was permitted? And remember the parallels to binge eating disorder and addiction?

But most importantly of all, it is easy to forget in the maelstrom of moral panic about obesity that food can be one of life's great pleasures. There is a point when we should stand up for joy. When public health campaigners insist that everyone eats a certain way or decry food choices as irresponsible and foolish, this is what they fail to understand. Food is not just a vehicle for nutrients, it is how we signal our identity to the world. It forms a vital part of all religions and cultures, both in terms of what people exclude and which items are eaten during festivals and events. Dishes bond people across countries and generations. The taste of something familiar can make us feel as if we are home, even when we are a thousand miles away. It enhances our lives in beautiful and innocent ways. The way that we eat is part of who we are.

Indeed, identity is sometimes more important than pleasure. A sugar-sweetened drink is mostly drunk because it tells others a little bit about us. Few consider Coca-Cola the greatest tasting beverage the world has ever known. No one thinks that energy drinks such as Red Bull or Relentless are irresistible taste sensations. They are an accessory, almost exclusively drunk in public, their distinctive branding visible to the world. Unlike many food products, they resist inroads from supermarket own-label options. Even if Asda managed to exactly copy the recipe for Diet Coke, something not beyond the wit of a talented product developer, their own label equivalent would not sell anywhere near as much, even if it was half the price. So if you are serious about behaviour change, this must be understood. It is no use saying 'if you are thirsty, drink water', because water does not satisfy the same thirst.

FOOD DESERTS

There is a common assertion that so-called food deserts, where access to healthier foods are limited within certain neighbourhoods, might be a driver of poor choices. Certainly, this is a compelling idea, and in the US it has received a great deal of attention. The elimination of food deserts formed a key part of Michele Obama's 'Let's Move' initiative,

designed to help tackle rising rates of childhood obesity. Evidence for
this approach centred around the observation that in areas where takeaways and convenience stores outnumber large supermarkets, rates of
obesity seem to be higher.[9] Of course, this association might only be
showing that in areas where obesity is prevalent there is little demand
for healthier foods. But as with most environmental factors, the relationship is probably two-way. It may well be that increased availability
of healthy food would help drive consumption and encourage better
choices, but it is probably equally true that consumer demand shapes
the retail environment. Businesses only thrive where the market exists,
and people who really want to eat healthily will probably be prepared
to walk an extra mile.

Perhaps because of this, evidence for the power of food deserts to
shape people's choices is inconsistent. Some US studies have found
that exposure to different food outlets has no influence on the diet or
obesity risk of school-age children.[10] In the UK, the introduction of a
large supermarket into a poorly served area of Leeds seemed to show
marginally improved consumption of fruits and vegetables, but a similar study in Glasgow showed no such effect.[11] In the UK generally,
there seems to be little association between the existence of food
deserts and fruit and vegetable consumption,[12] prompting the authors
of a 2005 report to suggest that instead of focusing on changes to the
environment, 'food policies aimed at improving diet should be orien-
tated towards changing socio-cultural attitudes towards food'.

It does appear that food deserts are strongly associated with
deprived neighbourhoods and poor health outcomes throughout North
America.[13] But in other developed countries, attitudes rather than
environment seem to dominate people's choices. Even in the US, the
relationship is only one of association. It may well be that American
planning regulations, land prices and more pronounced segregation
along racial and socio-economic lines cause food deserts to have a
greater impact.

The case against takeouts and takeaways seems to be stronger, and
this is certainly reflected in the actions taken against them. Analysis of
takeaway foods show that they are 65 per cent more energy dense than
the rest of people's diets and tend to be served in larger portions.[14]
People who consume large numbers of takeaways tend to have higher
BMIs, even when other factors are accounted for,[15] suggesting that
takeaways might well be a driver of obesity. Good correlations have

been found between the number of takeaways in a local area, the number of takeaway meals people consume and how fat people are.[16] It does seem likely that a high volume of takeaway outlets might be a good definition of an obesogenic environment.

In the UK, this has led to action. In 2017, the London Mayor Sadiq Kahn announced a complete ban on the opening of new takeaways within 400 yards of schools, claiming that this would help tackle the 'ticking time bomb' of childhood obesity. Twenty other UK councils have regulations in place prohibiting the opening of fast food outlets near schools.

Although it might seem logical that fewer takeaway outlets will lead to reduced consumption, it is troubling that these interventions have been introduced without properly evaluating the potential impact. They target a small group of food sellers, which does not include many of the large, multinational chains that generally operate under restaurant licences. It does not include bakeries selling pies and sausage rolls, or coffee shops with their high-calorie frappuccinos and cakes. The takeaways being targeted are fried chicken shops and kebab houses, the sort of establishments that serve working class people. Singling them out seems to favour businesses that appeal to more middle class sensibilities, with little justification as to why these are better. It is hard not to think that the motivation is more about cosmetically improving certain areas, driving out cheap looking locally owned businesses, and welcoming in the global players.

Some have suggested an even darker motivation. Dr Megan Blake is a senior lecturer in geography at the University of Sheffield, specialising in food security and food justice. She told me:

> Exclusion zones are a colossal mistake. Kids don't just get their food from next to school. It is a silly argument, and a blunt, ineffective tool. For the people who run these businesses, this is their livelihood. They are mostly Pakistanis and Bangladeshis, making it an institutionally racist policy. It is a shame, because there might actually be some potentially useful interventions you could make. Business rates for fast food restaurants are currently the same as for fruit and veg stores, so it is currently far more profitable to open a takeaway.

And despite some convincing UK studies looking at the impact of takeaways, there is a good deal of conflicting evidence as to whether or not the number in a local area has an impact on obesity.[17] This suggests that the relationship is highly individualised, with much depending upon the particular community. Megan Blake describes some of the reasons for this, perhaps giving a wider insight into why the environmental account of obesity is so complex:

> There is something to it [the environmental account], but not enough thoughtful research has been done yet. Why should we ban takeaway restaurants? Takeaways are not inherently bad because they are takeaways. Fast food and street food can be fantastically healthy. Interventions should be about improvement, creating more social eating occasions for people to eat together, and making healthy eating more pleasurable.

Given how mobile and flexible our society is, to restrict the areas in which takeaways are allowed seems foolish and potentially driven by distaste of the clientele. Takeaways fill a need in people's lives, especially those of young people, providing them with meeting places and forming part of their identity. It would be more progressive to engage and understand, rather than condemn and legislate. To improve what they are doing, rather than just insisting that they stop, perhaps starting by educating those still using hydrogenated oils in their fryers. Even if there was good evidence that reducing the number of takeaways will make people stop using them, to justify a ban, we would need to understand where they might be driven to instead. I doubt they would just have a glass of water.

AGENCY

In all this, we do need to consider what we are implying when we blame the environment for obesity, because we all occupy the same environment. Obesity has increased for everyone. Politicians, judges, bankers, construction workers, cleaners, chefs, students and doctors. Children, adults, men and women. While there are differing rates within each group, all have seen similar increases.[18] So when we say that an obesogenic environment has caused people to gain weight, and that changes to this environment will cure them, it does suggest that the fat people in each group are more susceptible than those who remain thin. The environmental account suggests that fat people are making a poor set of choices within that environment, and have less agency in their decisions than those who don't gain weight.

When it comes to making people alter their behaviour, the modern liberal suggestion is that these manipulations should take place behind the scenes – gentle behavioural nudges towards better choices. Yet rarely does this come with a definition of what 'better' looks like. Mostly, the idea is that fat people need to be coerced into more middle class food options. Cooking from scratch, meals round the table, less fried chicken, more hummus. There seems to be little evidence that this will help, yet still we push on, because something must be done.

Anna Kirkland, professor of women's studies at the University of Michigan, has been one of a small group of people who have questioned the environmental account of obesity. Despite it being the prevailing liberal view, it does raise some troubling questions. In a 2011 paper, she asked: 'What if it is the case that many elites find the terms of the environmental account to be simply a more palatable way to express their disgust at fat people, the tacky, low-class foods they eat, and the indolent ways they spend their time?' She went on to suggest that 'proper practices of food, eating, and exercise have been raised to the status of absolutely correct rules for good health rather than simple features of human cultural variety . . . There is a highly specific and evolved set of social rules governing the hierarchy of foods. A baguette is not junk food, but sliced white bread is; the sugar in honey and fruits is healthy while white granular sugar is junk.'[19]

The suggestion that fat people can't walk past a takeaway implies that they cannot be trusted. Normal weight people need to save these irresponsible sinners from themselves, and create an environment in which they will not be so damaged. Often the environmental account is little more than a thin disguise for contempt. Fat people are just slaves to the world around them, thoughtlessly shovelling in pies because food companies told them to. Of course, I do not think any of the experts I spoke to are covertly trying to manipulate their arguments to enforce norms of food consumption upon people. But environmental arguments for obesity have a real danger of treading that path.

There is always a temptation to try to change things that are easy, rather than address what is really broken. While the political right shamefully uses prejudices against fat people for political gain, here the left can be seen doing exactly the same. The liberal account of obesity presents it as a failure of the free market, with corporations left to run amok. The solution becomes an authoritarian control over the food supply. The choices deemed ideal inevitably mirror the food consumed by a rich elite, who assume that the errant fat people only gained weight because they had too much freedom.

Whenever we consider environmental behaviour change, we need to think about exactly what our end game looks like. Do we know what the perfect environment is, and are we really heading there? More than anything, the question we really need to ask is whether the new world we are creating is one that we want to live in.

BEYOND THE BIG TWO

There are two factors we know about already: eat less and move more. These provide us with simple, blunt and ineffective solutions to the problem of obesity. Whether you call for more personal responsibility, or demand dramatic changes to the environment, you are still reducing a complex problem down to a simple trope. But what if there is something else we are missing? What if there are other simple ideas that might prove valuable for our understanding of obesity? There are four other factors that are worth discussing, the first of which is the large chunk of our lives in which we're neither eating nor moving.

1. Sleep

We strongly associate fatness with laziness and so the idea that a lack of rest might make us heavier is not easy for us to comprehend. It does not help that our culture has a machismo connected to poor sleep patterns. Not getting enough is often used as a signaller of status, a badge of honour used to denote how valuable our time is.

One of the lasting memories that many people have of conservative Prime Minister Margaret Thatcher is from the aftermath of the 1984 Brighton Grand Hotel bombing. After a devastating explosion ripped through her hotel at 2.54 a.m., Thatcher walked out unscathed, fully clothed, hair and make-up completely intact. Despite the late hour, she had been awake in her room and working on her conference speech. This was lasting proof that the apocryphal tales of her ability to function on little or no sleep were completely true.

Thatcher famously got by on only three or four hours of sleep a night, a trait supposedly shared by Churchill during the war years, and claimed by a number of high achieving individuals. It now appears that her ability to do this might have been dependent upon a specific genetic adaptation present in only a small percentage of the population,[20] but at the time it was attributed to her extraordinary dedication and focus. This fitted nicely into the dominant neoliberal ideology of the time. If you want to be successful, you need to work hard. If you want to be really successful like Thatcher, you have to go without sleep. You snooze, you lose.

In the modern day, we often hear that many of us are not sleeping well, and it may be doing us a good deal of harm. Cultural changes have been exacerbated by the continued growth of artificial light and an increasing use of video screens, both thought to disrupt sleep patterns. But despite it being commonly reported, there is very little good evidence that we are sleeping less than we used to. A 2012 systematic review of data from a number of countries, including the US and the UK, found no good evidence that the average amount of sleep has decreased at all since the 1960s.[21] Perhaps even more surprising, studies of hunter-gatherer tribes in Tanzania, Bolivia and Namibia found that they averaged considerably less sleep than people in modern industrialised societies.[22]

This suggests that lack of sleep is not uniquely responsible for modern rises in obesity. But on closer inspection, there seems to be something more interesting going on. Although the *average* amount of sleep has changed little, the number of people defined as short sleepers, those getting less than six hours a night, has risen significantly. In 1975, 7.6 per cent of people reported being short sleepers, but this rose to nearly 12 per cent by 1999, before falling to 9.3 per cent by 2006.[23] The rise was particularly significant among full time workers, with students apparently getting more sleep these days (perhaps not much of a surprise to anyone who knows one). Short sleepers work longer hours and dedicate more time to personal leisure activities. They work hard, play hard and they'll sleep when they're dead.

Not getting enough sleep might seem a triviality in our busy modern lives, but there is pretty good evidence that it is anything but. Much of our functioning is dependent upon circadian rhythms, fluctuations that occur daily, largely in response to the cycles of light and dark outside our windows. Crucially, circadian rhythms are known to have an effect on the hormones that govern our appetite. Most of us are quite happy to go without food for ten to twelve hours during the night time, but would suffer hugely if we tried to do the same in daylight hours. Leptin, ghrelin, insulin, peptide YY and cortisol are all known to vary in circadian patterns throughout the day,[24] and disruption of this rhythm is known to have significant effects. Although the mechanisms are not completely clear, when our environment or behaviour disrupts our circadian patterns, it can lead to metabolic disorder and worse. Shift working is listed as an activity that is 'probably carcinogenic to humans' by the World Health Organization, as is working as a flight attendant.

Association between sleep patterns and obesity are common and easy to find. Research suggests that poor sleep has been shown to alter ghrelin and leptin levels, increase hunger and make people more likely to choose calorie dense foods. The amount of sleep people get seems to be associated with obesity in a U-shaped curve, increasing sharply for anyone getting less than six hours, falling to its lowest for those sleeping between six and eight, and increasing for anyone sleeping longer than that (students, you can stop looking smug now).[25] Although the *average* amount of sleep people are getting has changed little over the years, a shift towards the extremes of this U could have created significant problems. Short sleep duration in children as young as three has been found to be a strong predictor of future weight gain,[26] and some have suggested that the relationship between increased screen time and obesity may have more to do with sleep disruption than a lack of activity.[27]

'Social jetlag' is a term used to describe the regular disruption of circadian rhythms due to lifestyle and work patterns. When our inbuilt biological clocks run out of sync with our lifestyle, our body reacts in a number of negative ways. It is associated with a higher BMI,[28] impaired glucose tolerance and even lower life expectancy, suggesting that those who neglect sleep in favour of other temptations might be putting themselves at significant risk.

Certainly the evidence for a lack of sleep being a driver of obesity is significant and much stronger than for any specific type of food. Although most of the studies only show associations,[29] there is also plenty of evidence of mechanisms that might drive metabolic problems.

Yet sleep is almost entirely overlooked in the obesity debate in favour of the dominant food and exercise arguments. Although there has been some recent focus on the general health benefits of sleeping well, it is curious how the demonisation of supposedly irresponsible food choices has not extended to staying up late watching Netflix. Lack of sleep means hard work, socialising, activities, games and early morning exercise. It is dinner parties, movie nights, and other things that the elites of this world are not willing to sacrifice.

To be clear, I am not suggesting we also demonise those who do not sleep well. People should be able to live whatever life they choose, so long as they are not directly harming others. For many, shift work is not a choice, and they are forced into broken circadian rhythms. Shift workers are our doctors, our police, our carers, and they keep the world running while the rest of us sleep. Although society should make an effort to avoid shift work whenever possible, we should also make sure we thank those who make the sacrifice.

Given the good quality evidence that exists for the importance of sleep, and the damage that a lack of it does to our metabolism, it is strange that it is rarely built into public health strategies designed to combat obesity. It is such a simple thing.

2. Cold

What else in our environment might be having an effect? How about heat? Depending upon how cold it is, a considerable proportion of our daily calorie expenditure comes from keeping our body temperature constant. The colder we get, the more energy it takes to keep us warm. Perhaps as things heat up, the calories saved might get stored as fat.

Clearly, even if we take into account global warming, it seems unlikely that global temperatures have changed dramatically enough to affect people's body weight. But how about our climate-controlled homes, cars, public transport and offices? Many people looking back throughout their lives will remember a time when it was not uncommon to be uncomfortably cold at home during the winter, to shiver yourself to sleep, and to pile on multiple layers to keep the thermostat turned down. In our modern world, many more of us have money to spare, our homes are better insulated and our heating systems more efficient. Maybe we are fat because we never allow ourselves to get cold.

For once, our perceptions of things being different 'back in the day' might not be far from the truth. Between 1970 and 2000, the average temperature inside UK homes increased by 5 degrees centigrade, rising from 13 to 18 (about 55 to 64 Fahrenheit). Experiments indicate that this sort of difference can result in a decreased metabolic rate of around 170 calories a day,[30] meaning that higher temperatures result in us burning off significantly fewer calories. If even part of this spare energy was converted into body fat, it would result in people becoming overweight extremely quickly. Such a difference would be more than enough to explain the UK's increase in obesity levels over the past fifty years. This might seem far-fetched, but rats housed at 25 degrees centigrade store significantly more fat than those kept at 18 degrees, and the colder rats also manage to lose excess weight at a much faster rate.[31]

Although a tempting proposition, if it genuinely played out, we might expect that warmer climates would be strongly associated with obesity, which does not seem to be the case. It seems likely that any change in energy expenditure is compensated for, presumably by us eating less. Anyone who has ever experienced a hot climate will attest that it significantly decreases your appetite, perhaps because your energy needs drop substantially.

But even this relationship is complex, and perhaps dependent upon the physiology and genetics of individuals. Studies of African American and Puerto Rican children show that those born during colder months of the year are more likely to gain weight during childhood, perhaps indicating that laying down fat stores in response to cold weather might be a protective adaptation. And consistent exposure to very high temperatures is also known to increase our metabolism, as we are forced to use energy to bring our body temperature down.[32]

However, it is well worth noting the changes caused by temperature if we want to understand the rises in obesity. For many of us it is likely that our calorie requirements are lower than they were fifty years ago due to increasingly warm, comfortable lives. This might partly explain why some populations seem to have gained weight despite food intake remaining unchanged.

There is also another factor affected by heat that we are only just beginning to understand. It appears that the temperatures that we are regularly exposed to might affect the fat we have in storage in a way that could have a profound influence on our health. There are thought to be a number of distinct types of fat cells performing various functions in the body. White Adipose Tissue (WAT) and Brown Adipose Tissue (BAT) have different roles. Many think that increasing levels of brown fat might one day lead to novel treatments for obese patients, making this a hot area of study.

Infants have high levels of BAT, which are known to have a role in generating heat and keeping their young bodies warm, but the amount diminishes quickly as we age. Until relatively recently it was thought that the vast majority of adult fat was white adipose tissue, yet it has recently been discovered that we retain significant amounts of brown fat, which remains inactive until we are exposed to the cold. As recently as 2010, it was discovered that white fat cells could partly convert into brown fat cells, creating a third type, known as beige fat. Various environmental factors are thought to be involved in this process, including, unsurprisingly, drops in temperature. Perhaps low levels of beige fat caused by our climate-controlled lifestyles might provide an additional mechanism for temperature to affect obesity.

There are also a few dietary factors that might increase the production of beige fat cells, including capsicum, the hot compound in chillies, and certain chemicals found in olive oil. But interestingly, changes to stomach bile acids and alterations in the gut microbiome also seem to have an effect, something that we shall return to when we discuss weight loss surgery in chapter 17.[33]

Brown and beige adipose tissue helps with a process called non-shivering thermogenesis, generating heat from fats and sugars rather than converting it for storage. A word of caution: increasing heat production in fat cells might sound like a great way to burn off calories for free, keeping you toasty and warm in the process, but it comes with potential dangers. An illegal weight-loss supplement known as DNP activates similar pathways, and has resulted in a number of people literally cooking themselves to death from the inside, as their bodies convert too much of their food into heat.

Strangely, the conversion of food into heat is not thought to be the main reason why these different fats might be protective against metabolic disease. Brown and beige fats are thought to signal to the metabolism in a particular way, producing different hormones and factors that might help to reduce inflammation, even when they are located in our visceral fat stores. Unlocking exactly how this works will be key to understanding how it might benefit us in the future.[34]

3. *Maternal age*

We previously discussed whether people having children later in life might have an effect due to assortative mating. Between 1984 and 1994, the mean age of UK mothers at childbirth increased by 1.4 years, and there have been broadly similar increases in the US and Canada. In addition, the number of women giving birth aged over thirty has reached record figures in recent years. Between 1990 and 2004, the number of births to women between thirty-five and thirty-nine increased by 50 per cent. In the same period, the births to women over forty more than doubled.

It is a little known fact that having an older mother has been shown to increase the chances of becoming obese, a relationship that has been shown in numerous studies, some dating back to the 1970s. Older mothers are known to have an increased risk of bearing children of very high or very low birth weights, and both of these events are known to be risk factors for future obesity. In addition, older women are more likely to suffer from a number of health conditions including pre-eclampsia and hypertension, known to adversely affect the foetal environment, and have an influence on childhood weight gain.[35]

It is estimated that the increases between 1984 and 1994 raised the odds of children becoming obese by around 7 per cent. But here is a perfect example of a risk factor that we know about, but would probably not wish to change. Women giving birth to children later in life is a proud marker of our progress as a society. No one wants to return to a time when a young girl's lot was to marry young and procreate. Later childbirth has allowed many more women to have careers, to shape the world, to be educated, to engage in scientific research, to become leaders and politicians. It is hard to think of a development that has done more in raising levels of equality and standards of living. It has enabled economies to develop and thrive in ways that could barely have been predicted sixty years ago.

Sometimes great things come at a cost that we were not expecting. But often this is a price worth paying.

4. *Endocrine disruptors*

I am not a fan of spreading fear that our modern industrial environment might be toxic, and I certainly don't believe that we are being constantly bombarded by dangerous chemicals. In particular, I do not think we need to engage in spirulina and wheatgrass cleansing rituals to eliminate our 'toxic load'. However, there is a plausible argument that we do find certain endocrine disruptors in our modern environment.

Endocrine disruptors are a variety of industrially produced substances that can disrupt the functioning of our hormonal systems, with the potential to affect how we regulate our body weight. Many are thought to exert a particular influence on oestrogen receptors, which are well known to affect fat storage. There is good evidence that these endocrine disruptors have increased in the food chain in the past sixty years, and one study even showed that concentrations are sometimes higher in people with more body fat. They are resistant to being metabolised, and can become stored within fat cells, often passing up the food chain when animal products are consumed. For instance, phthalate is a chemical that has been shown to cause endocrine disruption in animal experiments, and it is estimated that up to 75 per cent of the US population have detectable traces of it in their urine. It has also been shown that presence of these traces is significantly correlated with obesity in human populations.[36]

I am not about to start calling for an intense moral panic about deadly fattening chemicals in our environment, but it does appear that there is cause for more investigation. Obviously the presence of phthalate, or any of these chemicals, might be at levels that are completely harmless. The fact that we can observe them might only be testament to our ability to detect vanishingly tiny amounts, way too small to cause metabolic harm. And the correlations with obesity could well be occurring because contamination is largely from animal fats, and so higher levels simply reflect the amount of these products being eaten.

But endocrine disruptors have the potential to affect our metabolism when present in extremely small quantities, and in areas where contamination is prevalent they might well be causing some harm. We have done so much research into obesity, and yet there is still so much that is unknown.

ANYTHING ELSE?

There are a few other factors, particularly those affecting children in utero, but these are complex and beyond the scope of this book. Certainly, the factors that create an obesogenic environment are many and varied, and are not always a matter of personal choice. They are interconnected, pushing and pulling at each other in different ways, perhaps only having a noticeable effect in combination.

I do not suppose anyone has all the answers, or has even asked all the right questions yet. But one thing is for sure: the picture is far more complex than most people think it is and many important factors are frequently left out of the discussion. Even here, I have barely scratched the surface, and a full exploration of each topic in this chapter would easily fill a book in itself.

I have repeatedly said that BMI has increased broadly across many different groups. This is a vital point, the importance of which many people fail to fully comprehend. It shows that whatever environmental factors are causing BMIs to rise and skew to the right, they are affecting all of us, not just a few specific people in troubled environments. We are all getting fatter, and nothing is insulating us from obesity's rise.

In recent years, however, there has been one specific group that seems to be at an increasing risk of getting fat. In the next chapter I want to examine this, and look closely at the reasons that might underlie these increases in the hope of revealing a few more clues as to what is really going on.

16

IS IT BECAUSE WE
ARE POOR?

Don't overlook the money part of it. I've been rich and I've been poor. Rich is better.

Beatrice Kaufmann

It is a strange paradox of modern industrial societies that if you are poor, you are more likely to be fat. For so long, corpulence signalled a rich elite, shamelessly overconsuming, while those around them struggled to feed themselves. Perhaps this explains some of society's disgust when it comes to obesity: for much of history, to be fat was to take more than your share.

But things have changed. Professor Sir Michael Marmot is an epidemiologist and director of the University College London Centre for Health Equity. He has spent many years studying how income inequality affects people's health, something eloquently described in his 2015 book *The Health Gap*. He told me:

> In low income countries, obesity is associated with high status and education levels. Only the top earners are likely to be obese, a trend particularly seen in women. But when incomes rise to levels above $2500, an inverse association starts to be seen. In high income countries, where everyone has enough calories, low income and status becomes associated with obesity, with the patterns being much clearer for women. Education can be protective. In Egypt, for women in rural areas, increased income

is associated with increased obesity. But for urban women with higher levels of education, there is not the same pattern.

Increasingly, particularly in countries such as the UK and the US, being fat has become an outward sign of being poor. And things only seem to be getting worse. Recently, as numerous campaigns and initiatives have attempted to curb the increases, some success has been seen. But even this hides troubling inequalities. Michael Marmot explains that 'there are huge inequalities. The rise in childhood obesity has levelled off for affluent children, but not for the least affluent. The more deprived children are, the greater the rise in obesity rates.' When I ask him why this might be occurring, a problem that has long troubled me, like any good epidemiologist he refuses to be drawn into definite causes. But he does provide an interesting insight: 'It is related to the way the social environment is interacting with the mind. A narrative about social responsibility doesn't account for these trends, because then you are saying that people are irresponsible on a social gradient.'

Neoliberal ideology is quick to blame obesity on the bad behaviour of individuals. The implication is that poor people, and increasingly poor children, behave less responsibly than their rich counterparts. Do we really think people living in poverty inherently lack willpower? Or are we all products of our circumstance, playing the cards life has dealt us as best we can?

There are many other questions here, none with simple answers. Does being poor make you fat? Conversely, given the stigma and employment prejudice we have already discussed, does being fat make you poor? Does poverty lead to poor food choices, and if so, why? Is there something about poverty beyond food that makes people gain weight? And if the link is so strong, how might it be broken?

But perhaps the most significant question of all, why is it that these problems are increasing? We have always had poverty, and for a significant part of the twentieth century, most people in developed countries have had access to enough calories. On average, diets have improved in the past forty years, not worsened. What is it about the modern-day experience of poverty that makes people fat? And why did working class communities in the 1960s and '70s not suffer from the same problems?

This is the final chapter of this part, and likely to be the most divisive. Poverty, and the vast health inequality that results from it, is something that shames us all. Most of us would rather not consider the truth of the matter.

POOR WOMEN

In the UK, the prevalence of obesity in the least socially deprived adults is a little over 21 per cent. For the most deprived, prevalence is just over 29 per cent, perhaps not as dramatic a difference as some might think. As Professor Marmot suggested, however, the difference is considerably starker when we look at women alone. Women in the least deprived areas are obese with almost exactly the same prevalence as men, 21.7 per cent. But in the most deprived areas, the prevalence is nearly 35 per cent. For some reason, women seem considerably more exposed to any socio-economic effects that might cause them to gain weight. Troublingly, women are also considerably more likely to become severely obese. There are over twice as many women with a BMI over 40, at around 3.5 per cent of the population. These statistics present an obstacle to the account that an obesogenic environment is solely responsible for obesity.

The picture is worse when we look at UK children. Setting aside the measuring problems already discussed, at the age of ten, about 11.5 per cent of children in the least deprived areas of the UK are obese. At the same age, the rate is 25 per cent for children in the most deprived areas, over twice as many. Even more concerning, between 2007 and 2015, the percentage of obese children in the most deprived areas rose every year, climbing three percentage points over the period. For the least deprived, the number fell consistently, dropping one percentage point. The gap is huge and widening.

Perhaps this is not surprising. Interventions are usually designed by the political classes, or public health workers leading isolated lives. They are created by the elite, and perhaps that is why they only ever reach the elite. Eatwell Guides, Sugar Smart apps and Change4Life recipe cards have no relevance to many people's lives, and can often seem patronising and distant. Maybe it is also because most of the help being offered comes from top-down interventions. We are told what to do, and will only comply if we feel the need to be good citizens. It is only the involved and engaged who answer the call. For the most denigrated members of society, those who are consistently told they are of no worth, what's on offer seems of little relevance. When you are struggling to make ends meet, with debt-collecting wolves at the door, being a good citizen is pretty low down on the list of shit to worry about.

SORRY, HULL

A couple of years ago I wrote an essay entitled 'Because it's Fucking Shit in Hull', which discussed how the experience of stress affects long-term decision making. It was based on a discussion with Professor Michele Belot, an economist at Edinburgh University. She speaks passionately and eloquently about how people in stressful situations are more likely to do something called 'delay discounting'. This makes them more likely to go for the short-term win than hold off for a potential long-term benefit. If someone has to choose between £10 now, or £20 in a few weeks' time, those inclined to discount delay will be more likely to take the immediate option. Then, they will probably head straight for the pub.

High levels of delay discounting have been shown to have significant deleterious effects on people's lives. It makes them more likely to smoke, more likely to eat unhealthily, inclined to risk and prone to engage in criminal activities. They will be less likely to save money, spend less time in higher education and rarely learn any complex skills. They will earn less in their lifetime, die younger, develop more chronic illnesses and have more serious accidents.

How much people delay discount can vary enormously with circumstance. For instance, if someone is dying of thirst, a drink of water right now will far outweigh any amount of money further down the line. But it has also been shown to have a significant correlation with poverty. Michele had suggested to me that one of the strongest theories explaining this connection is that people living in poverty simply have more stuff to worry about than those born into privilege and wealth. A recent Harvard study investigated the ability of poverty to consume people's attention and create a hierarchy of worry, with healthy food choices falling very much into a second tier of concern.

I have never been completely comfortable with the idea that delay discounting makes poor people more likely to be fat. In my essay, I suggested that far more of the supposedly bad decision making might be due to a misplaced sense of identity. The idea that in certain communities, any ambition to transcend your birthplace through pious self-control and engagement in education is frowned upon or mocked. This was neatly summed up in a comment I received after posting the essay: 'It's hard to tell your mates that you want a better life.' I would add that for many it might be equally hard to tell your family. If you have lived in a working-class community, among people in low-paid jobs who live for occasional carefree nights out, telling them that sort of life is not enough for you is hard. Investing in yourself in order to rise out of your community can feel like slapping the people you love in the face.

And so, deprived communities build poor long-term decision making into their identity. In many cases it becomes framed as a positive asset. People live for the now, don't care about the future and enjoy the moment. Those who delay gratification are boring, stuck up and pretentious. When poverty is so engrained into the identity of a community, discounting delay becomes part of that identity. That's just what you do round here.

So when a young mother living on a rough estate in a deprived town in Northern England was asked why she did not have any ambition to better herself, she replied, 'Because it's fucking shit in Hull, no one from here ever does anything.' She felt that her community was so broken that it made ambition pointless.

Yet for a young, aspirational middle-class woman, ambition is very much part of her identity. She will make sacrifices to get what she wants, because that is what people like her do. Both identities are accidents of birth, yet with very different outcomes.

I worked for many years as a chef in restaurants and hotels in Manchester. Chefs do not earn much money, and I lived in some of the poorest and most run-down communities in the city. My friends and colleagues were all hardworking and low paid. We lived for the moment, had wild nights out and rarely planned for the future. Our lives were all about the here and now, and we completely discounted the benefits of any future gains.

Far from being permanently stressful, it was fun. Even though I had responsibilities and struggles, I look back on it with fondness. I certainly do not remember a level of constant stress, and reject that I lived in the present just because times were hard. It was just my identity at the time, and what I wanted to do. Chefs are notorious for hard living – I was simply fitting into that mould. For this reason, I have always rejected any idea that people in poverty are under significantly more stress than those who are not. Certainly, it is not easy, but whose life is? There are good times, pleasures and happiness to be had even when you are poor.

But the more I have thought about this, the more I think there is a chance that I was insulated from stress for two important reasons. And within these is a big clue as to why more and more people are struggling.

BEING BROKE AND BEING POOR

In her astonishing book *Hunger*, Roxanne Gay describes the significant difference between being broke and being poor. I have been broke many times, but because of other aspects of my life, I have never really been poor. Because of my family, upbringing and education, all of which are distinctly middle class, I have always had a safety net. I have always known that if I fell, I would be caught. I have skills, knowledge and social connections that make it easier for me to get out of trouble. And I have never had any feeling that success would somehow betray, or distance me from, my upbringing.

So even though I have lived without money, failed to pay my rent, even gone without food at times, I do not know what it is like to be poor. I have lived in some of the most deprived communities in the UK, and counted some of the lowest paid and socially denigrated people among my friends, but I have never truly known a life of poverty.

Although my experience tells me that poverty does not always equate to stress, perhaps I am just mistaking personal anecdote for evidence. It may well be that stress plays a far bigger part in the experience of poverty within an unequal society than I realise.

Darren McGarvey is a Scottish writer and recording artist who grew up in Pollok, a deprived area on the South Side of Glasgow. In his book *Poverty Safari*, he explores the reasons why British working class communities have been so left behind by the rest of society. Crucially, Darren's upbringing did not provide him with the same sort of insulation from the effects of poverty that I was blessed with. His experience of stress in these communities was very different to mine. He told me:

People are not self-aware. They only know stress. They only know fight or flight. For people living in constant stress, life is like the dentist chair, with the drill forever in your mouth. We mistake stress management tactics for personality. You don't realise that everything you do is about stress. Affluent people have an emotional advantage. They use stress as a form of propulsion, rather than the soup they are swimming in. In many ways, the experiences and values of people are not different, it is just that for people not living in stress, it's simply something that has not occurred to them. When you braid all the experiences of poverty together – constant financial insecurity, threat of violence, lack of mobility and communities that look broken – some people will transcend it, but most people will be consumed.

Maybe for some of the people around me, the drinking, fighting, aggression, late nights, crap food and cheap drugs were just ways of managing difficult lives. Perhaps instead of being signals of a fun-loving identity, these were ways of blocking out the pain of a desperately tough existence. People lived for today, because the prospect of endless tomorrows was too much to bear.

I imagine the reality lies somewhere in between, but I realise I may well have underestimated the effect stress has on people's choices. There is, it seems, an intense relationship between stress and poverty. And there is also good evidence that this will create a tendency to become obese.

Stress is known to drive people to make poor dietary choices. Cortisol, a hormone released when people are under constant low-level stress, is known to drive cravings for sweet and fatty foods.[1] It has also been linked to an increase in snacking and a decrease in fruit and vegetable consumption. Cortisol is also known to change how your body responds to the food that you eat. When cortisol is in your veins, you are more likely to store energy as visceral fat around your organs. This in turn raises people's risk of developing metabolic syndrome, and many of the other diseases associated with obesity.[2] It has been known to have adverse health effects on children as young as eight, and is thought to be a contributing factor in the development of childhood obesity.

In short, stress can make you fat. And once you are fat, being stressed means you are more likely to get sick.

Of course, all this tells us is that poverty and stress are interlinked, and that stress can cause health problems, which is perhaps hardly surprising. But the experience of poverty is certainly not a new one, and has long been associated with hardship, insecurity and a threat of violence. Which makes me wonder, why is it that it has become associated with obesity only recently?

Here, it is perhaps worth considering the other aspect of my life in Manchester that I feel insulated me from the effects of my misspent youth.

TOGETHERNESS

Throughout this book I have talked about hunger and a desire to eat as being the primary human drive. Food is the thing that motivates us above all else, a desire to keep ourselves alive and fed. Along with breathing and drinking, it underpins all our other needs and desires. But increasingly it seems that people who study the human mind are realising that something else might be an even more significant force.

In many vertebrates, hunger rules above all else. But mammals have a problem, something particularly compounded in humans. As a consequence of evolving big brains, we are born helpless into the world. A screaming child is certainly hungry, but hunger is not its most significant drive. After all, food alone is useless to a child incapable of feeding itself. A need to be nurtured and cared for is by far the most important thing, trumping even its hunger. As a result, the strongest, most powerful forces within our brains are dedicated to the creation of social bonds. Babies are programmed to form connections with those around them, to ensure they are kept alive. And for the adult humans that child interacts with, the desire to care and nurture is by far the most powerful instinct of all.

Large parts of our brains are dedicated to the formation of social bonds, and the management of the complex connections we have with other people. When we are involved in cognitive tasks, neuroscientists can observe which part of our brains are involved in performing them. But when we stop considering the task in hand and move into our resting mode, other parts of our brains consistently light up. When we put our brains on stand-by, the regions that become active are those involved with the forming of social connections. Whenever we get a spare moment, we choose to work on the complex interactions we have with others. Although human brains can achieve incredible things with our amazing reasoning and logic, the reality is that we are social creatures at our heart.

This is why social isolation can be so devastating. It is why solitary confinement is the harshest punishment the judicial system has to offer. In neurological terms, our reaction to social rejection is exactly the same as our reaction to physical pain. Although we are taught to think of them as separate, it is no coincidence that in almost every language, the words used to describe social pain are the same as for its physical equivalent. Lost friendships hurt, hearts break and the loss of love cuts deep.[3]

When it comes to stress, our ability to manage it seems vastly improved by the formation of strong social bonds. For those who fight in wars, the effects of almost incomprehensible trauma are alleviated by the connections that form between comrades in arms. During sustained violent attacks on cities, levels of mental illness and suicide often decrease as adversity brings communities together. We find togetherness in hardship, and it makes us stronger than we can imagine.[4]

In recent years, our experience of community has changed. In the UK and the US, whether it is caused by neoliberalism or our distance from conflict, people are living lives that are increasingly alone. Five million Britons report having no real friends.[5] One in five rarely or never feel loved. Inhabitants of the UK are less likely to know their neighbours than in any other country in Europe. Over a million older people report being chronically lonely. In 2018, a survey for the mental health charity Time to Change found that two thirds of British people have no one to talk to about their problems.[6] In the US, the average number of close friends people have has fallen dramatically.[7] In 1985, the most common number was three. By 2004, it had dropped further: a staggering 25 per cent of people in the US reported having no close friends at all.

In deprived areas the problem is compounded. The destruction of traditional industries has led to increasingly fractured communities. The idea of rising with your class, for so long the central tenet of socialism, has changed into an aspiration to rise above it. People are encouraged to better themselves and leave behind their community, creating increasingly fractured lives. Darren McGarvey experienced profound changes the first time he encountered the more affluent side of Glasgow, discovering a freedom from stress he had never previously known:

The moment I realised there was another way of living, I became frustrated by the narrow-mindedness. Why are you fighting about this territory? There is nothing here. It is different to working class communities sixty, seventy years ago. There was poverty, but there was a sense of community, a sense of purpose. Deindustrialisation and the rise of neoliberalism has disconnected people from their community. Global businesses like Frankie and Benny's exist on the periphery of the community, rather than being part of it.

In contrast, as a chef in 1990s Manchester, I felt part of a community. Despite being poorly paid, my work had status. We dreamed of achieving the success of culinary superstars such as Raymond Blanc and Marco Pierre White, and were bonded by hardship and purpose – perhaps enough to distance us from poverty's worst effects. Community and social bonds have been in rapid decline in recent years. It is easy to blame the destruction of traditional industries, and this certainly played a part. But did those industries, full of hardship and horror, really represent a better life to which we should return? It is shocking that so many communities were left to rot when the old industries died, but it is equally shocking that no one thought to defend and fight for the workers who replaced them. In the service industries, call centres, bars, hotels, warehouses, shops, supermarkets and restaurants, the workers are damned and ignored by all. The right looks down on the poorly paid and tells them they should just work harder. The left, mourning a lost industrial past, complains that these jobs are not 'real work', largely because they are fragmented and hard to unionise. Both sides make people feel as if their lives are not good enough, that they should seek something better.

Everyone seems to have forgotten that these jobs are done by real people, and all are of value. Someone will always be required to clean our offices and hospitals. People will need to wash the dishes, sweep the floors and care for the elderly in their homes. Whatever social nirvana we dream of, we will still need people to cook burgers, empty bins, drive vans, stack shelves and work tills. They have been abandoned, and exist in a world that has stopped caring for them. They are forever told they should leave their jobs, which are not seen as adding the slightest value to society.

I have known good men and women who have spent their entire

working lives washing dishes, sweeping floors or cleaning toilets. They have raised families, paid a lifetime of taxes, and lived full and rich lives. Yet no one in politics represents them. Everyone insists their lives are broken. No one fights for them to be paid more. No one calls for them to be valued and respected. They are taught to be ashamed of what they do, and learn to distance themselves from their communities. The political classes care nothing for their lives, yet benefit from their services every single day.

All of modern politics works to destroy working-class communities. They are a problem to be fixed. The modern socialist movement has utterly failed to engage with low-paid working people because most are in industries that resist union representation. This failure is seen as the workers' fault, and so they are left behind with no one to speak on their behalf. They are increasingly, and unsurprisingly, distanced from politics. And crucially, they are distanced from each other. They reject the communities they live in. Many actively destroy the world around them. Their hopes and dreams are to leave their life behind.

With communities utterly broken, and people suffering alone, it is perhaps unsurprising that obesity has taken hold. People are unprotected from stress. This drives poor choices, and makes the consequence of those choices progressively worse. People who are isolated have less desire for discipline and the attainment of social norms. Stress is more likely to consume them, and when it does, they are far less likely to care. And for women, these problems are more intense, leading to higher rates of obesity and sickness. Women are less likely to be in work, denying them identity and status. They are more likely to be lone carers for children, with all the stress and responsibility that entails when living on a low income.[8] Single mothers routinely receive many times more admonishment than the fathers who callously abandon their own children. It is hardly surprising that rates of obesity are higher for women in poverty, and for the children who live with them.

As the connection between poverty and obesity has strengthened, fatness has become a signifier of social class. This has driven deeper prejudices and fears within affluent communities, with obesity being seen as a sign of moral failing. The rise of 'poverty porn' television in the early 2000s, where the lives of people on welfare became the subject of mainstream reality documentaries, has quickly transformed into 'fat poverty porn'. Titles such as Channel Five's *Too Fat to Work* and *Obese and on Benefits* leveraged our suspicions and prejudices against fat people, and neatly attached them to our feelings towards scroungers. In the watching public's eyes, the two facets of these people's lives are intimately connected. The *Daily Mail* runs endless stories of greedy fat people fraudulently claiming benefits or, God forbid, having their houses modified to help with their mobility. More and more, our society insists we isolate and revile the poor.[9]

The neoliberal dream has far-reaching consequences. That it turns us on our communities is unsurprising. The aspiration that you should rise above your class implies contempt. We are taught to blame people for their bad choices, seeing them as the cause rather than the result of an unequal society.

WHAT TO DO?

Endlessly we are told what should be done. How the environment should be shifted and shaped to make poor people engage in more middle-class choices and values. We should teach 'them' to cook, to stop them lazily sitting at home pressing microwave buttons. Surely when they know how to make a beetroot, smoked mackerel and quinoa salad from Yotam's new book, they will all be better behaved.

Stress creates a fierce demand for anything that might soothe it, if only temporarily. Drugs, alcohol or comforting foods, particularly the cheapest sources. This is what shapes the environment in which people live. You can regulate and tax to your heart's content, but you will not change this demand. Darren McGarvey explains:

> The local economy is all about providing stress relief. This develops into an addiction, leading to more stress. The political classes are useless, because they have no insight into the experience of poverty. You can modify the environment, but you really need to change the demand. Demand will always be supplied. If you stop the sale of high calorie foods, you will still be left with millions of people who are ill at ease. And they will turn to the next thing.

Even if we are allowed to rip apart their food supply, take away their choices and force them to exercise, will we really be improving their lives? They will still be poor. They will still be isolated. The constant stress they experience will still be pumping cortisol through their veins. All you will be doing is telling them that the decisions that make sense in their lives are foolish, and that they are not responsible enough to choose for themselves.

So go ahead, take away the chicken nuggets they were going to have for dinner tonight. The ones that they can afford and that they know their kids will eat. Replace them with something that you deem better, more acceptable, more middle class. See if that really transforms their lives.

What will happen is this. They will get on with it, because they are far more resourceful and capable than you might imagine. But all the stuff that makes their lives shit will still be there. The stress, the inequality, the loneliness. The stigma they experience every single day. The shame people pile upon them because of their bodies, their clothes, their manners and their appearance. The pervasive thought that they are a bad parent because they struggle to provide for their children. All that stuff will not have gone away. The only difference is that now they will be pissed off because they haven't got any chicken nuggets.

The fear and revulsion that we have for people who are poor and fat come from our unwillingness to accept that our own success is anything other than luck. We want to believe that our lives are the result of good choices, strong morals, intelligence and superiority. We assume that our food makes us who we are, rather than being a consequence of how we live. And so, on all sides of the political divide, we decide that if we can force our food values onto the poor, they will become like us.

Just as we cannot restructure food environments and expect demand to change, we cannot restructure food systems to alleviate health inequality. If we want to do that, we need to tackle poverty itself. We need to restructure our society. We need to become more equal, more accepting, more caring. We need to stitch our broken communities together and try to understand what common purpose looks like in a world without conflict.

The 2010 Marmot Review into health inequalities stated in its introduction that 'the link between social conditions and health is not a footnote to the "real" concerns with health, healthcare and unhealthy behaviours – it should be the real focus.' But the result was not a tackling of social inequality. All that it prompted was political pressure to try to influence people's food choices. Traffic-light labelling, sugar reductions and calorie cutting. This is known in the world of public health as 'lifestyle drift'. We take the path of least resistance, and it frequently leads us miles off course.

Instead of demonstrating against the Coca-Cola truck selling soft drinks at Christmas time, perhaps we should put our efforts into fighting against homelessness, protesting on behalf of the millions of proud people being forced to use food banks and ensuring that people are paid a better wage that allows them to live with dignity. The truth is that inequality, something in which we are all complicit, is to blame. Years of low taxation. A dismantling of the welfare system. The corrupt banks that have become too powerful to fail. The persistent underfunding of healthcare. Our insatiable demand for ultra-cheap goods and services. The old industrial communities that we allowed to wither and die.

People are getting sick because they are poor. But in accepting that, we must look at ourselves and ask if we are willing to share our privilege. To do this, we need to accept that it is only luck that sets us apart.

When I first heard Michael Marmot say that the only way to tackle the rising rates of obesity in poor communities was to end financial inequality, I thought he was foolish and misguided. I thought that surely with all that we know, we can take charge of communities, encourage better choices – basically coerce people into being healthier. That cooking classes, vegetable workshops, farmers' markets and cycle tracks could save people.

Change is possible. But it is not about changing the markers of poverty or sprinkling a few of the trappings of privilege on top of difficult lives. It is about creating substantial change, relieving the worst effects of stress, fighting stigma and building a better society for everyone. It is about catching people when they fall and creating fairer political movements that make sense in the modern world.

Sadly, these things are harder to achieve than teaching people to make their own hummus. But maybe the hard path is the one that we need to tread.

PART III

WHAT SHOULD WE DO?

There is no such thing as absolute certainty, but there is assurance sufficient for the purposes of human life.

John Stuart Mill

17

SURGERY AND DRUGS

The benefits of biomedical progress are obvious, clear, and power-
ful. The hazards are much less well appreciated.

Leon Kass

Given how much is now known about how our body regulates its appetite and energy expenditure, it might seem strange that no truly effective drug treatment exists to help combat obesity. Leptin was discovered in 1994 and led to an exponential increase in the understanding
of the processes that control how much we eat and how much we
weigh. So why is it that twenty-five years later, there is no useful pharmaceutical intervention available that makes people less fat?

THE DRUGS DON'T WORK

Drugs to control weight have a distinctly chequered history. For many
years, the most effective strategy involved the use of amphetamines.
Anyone who has ever taken one of the many illegal versions of these
drugs will know that they certainly decrease your appetite. But they do
not leave you satisfied, relaxed and full. The appetite reduction is more
of an anxious knot in your stomach, of the type that you might get if
you were being chased by an axe-wielding paleo dieter, angry about
your latest blog post. You don't exactly feel full, but you certainly aren't
thinking about a snack. Or at least, that's what I have been told.

Amphetamines are one of the clearest examples of why losing
weight is not the same thing as being healthy. Although the '30-Day

Crystal Meth Plan' might achieve cult success in certain parts of the modelling industry, I cannot imagine it catching on as the next Paltrow-endorsed health sensation. Side effects of amphetamine use include cognitive impairment, anxiety, psychosis, paranoia, aggression, insomnia, teeth grinding, irregular heartbeat, high blood pressure, constipation, cardiovascular disease and erectile dysfunction. And they are of course highly addictive, meaning that their medical use has been strongly regulated for some time.

Amphetamines interact with neurons in the brain that contain noradrenalin, in particular a group that responds to signals from the gut, received through the vagus nerve. Amphetamines cause these neurons to send signals to the hypothalamus that we are under extreme stress, reducing our appetite and increasing energy expenditure. Amphetamines produce our fight or flight responses, making us alert and ready for action, but without the fear that usually accompanies them. Unsurprisingly, however, placing the body on constant high alert comes at a cost to our health.

Amphetamines work extremely well when it comes to weight loss, but even the less addictive versions developed later as prescription weight-loss treatments had many of the same devastating side effects, making their clinical use unpopular.

In the 1970s, the drug fenfluramine was developed, another hunger suppressant, still affecting appetite regulation in the hypothalamus, but this time acting on a different set of pathways. Fenfluramine targets neurons that use serotonin, causing a decrease in appetite and a small amount of weight loss. It too has a number of unpleasant side effects, including diarrhoea, drowsiness and depression, so was never considered to be a useful treatment on its own.

But the different pathways that this drug used to decrease appetite, and the seemingly contrasting side effects to its amphetamine-based cousins, led researchers to look at combining fenfluramine with the less addictive amphetamine phentermine to create a combination therapy. The hope was that the two drugs would be more effective together, and that some of the side effects might cancel out. This so-called fen-phen treatment was first tested in 1992, and led to a very promising average 16 per cent loss of weight over a thirty-four-week study.[1] The ideal doses were found to be lower than when the drugs were used separately, and the side effects did seem to be alleviated, leading many to suggest that this had the potential to become a powerful tool in combating obesity.

By 1997, it was estimated that sixty million people had been prescribed fen-phen, but problems were starting to emerge. It appeared that the therapy carried with it an increased risk of hypertension, and, perhaps even more concerning, it seemed to be leading to permanent heart defects in a small number of patients. In 1997, fenfluramine was withdrawn,[2] leaving few remaining pharmaceutical options for weight loss.

Anything that could replace fen-phen would have been a blockbuster drug of epic proportions, but sadly, despite huge amounts of research time and resources being invested in the hunt, little of note has materialised. Lorcaserin, hailed as a 'holy grail' drug, has been shown not to increase the risk of serious heart problems in recent trials, but long-term weight loss is modest and side effects include hallucinations, suicidal thoughts and painful erections. Rimonabant targets the cannabinoid receptors that produce the well-known appetite increases reported by cannabis smokers, who often find eating a slice of buttered toast a near religious experience (I have heard such stories, but can confirm nothing). Although this approach showed promise, it was linked to suicidal tendencies, depression and other psychological problems, and has been largely withdrawn from use.[3]

It might seem curious that treatments for obesity often stem from insights into addictive drugs like cannabis and amphetamines, but it really shouldn't be. These drugs are enticing because they target some of the most primal parts of our brain, those that govern hunger, pleasure, fear and sexual desire. These are the things that truly motivate us, and they are almost always interlinked. To find a drug that affects appetite without affecting other primitive drives is almost oxymoronic.

Many drugs that modify appetite also profoundly alter sexual behaviour, as the two are strongly entwined. We have no desire to procreate when we are starving or fearful, and, conversely, if we start to artificially suppress hunger, it can lead to surprising results. In the 1990s, an appetite inhibitor called alpha-MSH showed much promise, but was hampered by its side effects. True, men ate less during treatments, but permanent erections made life hard, if you'll forgive the expression.

Almost all work on anti-obesity treatments has targeted appetite regulation, and almost all has led to unacceptable side effects. If you fool the body into thinking it is full, it will activate other desires. If you convince it that it is under stress, it will run itself into the ground. Hunger is at the heart of our existence. We mess with it at our peril.

There is a drug that bypasses these hunger and satiety systems, creating a slightly cruder intervention to help with weight loss. Orlistat, known by the brand name Xenical, lowers the absorption of fat from our guts, causing a significant amount to pass straight through us. It generally results in only modest weight loss,[4] and often creates urgent, explosive diarrhoea. It can also result in the need for dietary supplements, as the absorption of many important vitamins depends on getting enough fat in our diet. But its main problem, perhaps unsurprisingly, is compliance. The drug makes people fight their hunger, particularly any desire they might have for fatty foods. And hunger usually wins any battle in the long term, meaning people simply stop taking the pills.

Despite everything that medical science has achieved, and all the conditions that have been conquered over the years, if you want a drug that can help you lose weight, the choices seem to involve mental illness, heart disease, explosive fatty poo, boiling yourself to death or a permanent erection. But there is one treatment that is remarkably effective. And the most curious thing about it is, no one really knows why.

BARIATRIC SURGERY

Bariatric surgery was first performed in the mid-1950s, but due to a number of serious long-term complications, it wasn't widely used until the late 1970s. The idea of the surgery was to reduce the volume of the patient's stomach, preventing large quantities of food being eaten and absorbed. These initial surgeries were gastric bypasses, where the top part of the stomach is joined to a point in the small intestine, missing out a large part of the stomach and a small section of the gut. This creates a smaller usable stomach pouch and limits the amount of food someone can eat in one sitting. The two other main types of bariatric surgery are gastric banding, where an adjustable band is placed around the stomach, and sleeve gastrectomy, where some of the stomach is removed. By far the most common procedure is the original one, known formally as a Roux-en-Y gastric bypass, and accounting for around 70 per cent of all operations. On average, bypasses seem to lead to more weight loss than gastric bands, and although it is a more difficult and more time consuming procedure, there are generally fewer post-operative complications.

In the 1990s, the introduction of keyhole surgery revolutionised bariatric procedures, meaning that they could be performed using far smaller incisions. This, combined with rising incidence of severe obesity, led to an explosion in operations. In 1998, there were an estimated 13,000 performed worldwide. By 2008, this had risen to 220,000, and by 2013 it had reached nearly half a million.[5]

All the procedures are incredibly effective at promoting weight loss. Generally, surgery is only recommended for people with a BMI over 40, but it is often also performed on patients with a lower BMI if they have associated health problems. This is particularly the case for type 2 diabetics, who seem to benefit uniquely from this sort of treatment, leading many to call for a more formal lowering of the threshold for anyone with this condition.

Perhaps the most surprising thing about bariatric surgery is we don't really know why it works. Although it seems obvious that a physical reduction in stomach size will affect how much people eat, the real benefits seem to stem from changes in the hormones produced by patients' guts. Reversal of insulin resistance usually occurs incredibly quickly, with patients often able to stop taking diabetes medicine within a couple of days. These effects happen a long time before any fat is lost, and seem to be due to complex changes in a number of gut hormones, and their effect on the regulation of blood glucose. The regulation of bile acid levels in the stomach is also thought to play an important role, as are changes in the gut microbiome.[6][7]

These hormonal changes also seem to affect appetite control. Despite how it may feel, for most of us, the sensation that we are full occurs way before our stomach has reached its capacity. This is why after a delicious meal, although we might feel as if we can eat no more, we sometimes find room for dessert. We feel full not because our stomach is straining, but because our guts release hormones in response to food being eaten. But when a really nice looking cake comes along, we can override these feelings. Contrary to what many people claim, we do not have a second stomach especially for desserts.

When people undergo bariatric surgery, it appears that these hormonal signals are strengthened in response to food, leading to patients reaching a feeling of fullness far earlier than previously. Their guts send signals to their brain to tell them to stop eating, and so the amount they consume at each meal falls rapidly. Even more remarkably, fMRI scans have shown that people's brain reward pathways are less strongly activated by energy dense foods after surgery, indicating that it can actually affect food desires and preferences. Surgery does the very thing that years of pharmaceutical research have failed to achieve. It changes the signals being sent to our brain that regulate appetite. When hunger tells you to eat, there is little you can do to fight it, which is why most weight loss interventions are doomed to failure. But surgery profoundly alters how we experience hunger, and resets the body to defend a lower weight. And when that happens, fat loss is dramatic, and incredibly quick. The improvements to health can be quite remarkable.

The vast majority of bariatric surgery patients lose large amounts of weight in a short time. On average, patients lose 60 to 70 per cent of their excess weight, often with little effort. Many come to surgery after years of dieting and restriction and deeply troubled relationships with food. Afterwards, they see weight drop off almost magically and their lives can be transformed.

Post-surgery, over 75 per cent of patients see type 2 diabetes completely resolved. Nearly 80 per cent see significant reductions in hypertension. Metabolic syndrome is reversed in over 80 per cent of patients, and 95 per cent claim that the operation has improved their quality of life.[8][9] These dramatic improvements make bariatric surgery unique among medical treatments. It simultaneously treats a huge number of serious conditions, many of which are life threatening. This has led to claims that we really should be doing more of these procedures. The UK lags behind much of the developed world in rates of surgery, leading a large number of high-profile doctors to campaign for it to be used more freely.

As a dietitian working in weight management, Nichola Ludlam-Raine often works with patients undergoing bariatric procedures, helping them to adjust to the many changes that can occur. She told me:

> Surgery definitely has a place. For some patients, it is probably the only way to achieve long-term weight loss. I have worked with hundreds of patients, and in one or two it may not go to plan. But in most, there is a huge improvement, something that they would never have been able to achieve without surgical help. It can give people who are really struggling a lifeline. They go from having huge mobility problems, to being able to play with their kids. It can completely transform their health and

their relationship with food. People have told me that after surgery, for the first time ever, they know what it feels like to feel hungry and then full.

Sven Schubert is a patient advocate for bariatric surgery. He told me:

I was born prematurely and my mother was told I would be in hospital for six to eight weeks. After ten days, I had gained enough weight to be sent home, with the nurses joking that they had to get rid of me because I was feeding them empty. By the age of three I was wearing clothes meant for five to six year olds. Throughout my school years, I became so big that I struggled to play with other children. By nineteen I was 220 kilograms. By twenty-five I was 250 kilograms. At thirty, I was 300 kilograms, by which time I had stopped working.* I couldn't cope with life at that size. I broke chairs, broke floor tiles and couldn't get into my car. I had a gastric bypass at thirty-two and dropped from 300 kilograms to 145, and I maintained that weight for three to four years. I felt full for the first time ever. My body was saying 'it's okay, you can stop now'. The new feeling of being full was hugely positive.

Currently, in the UK at least, surgery is seen as a last resort, only used when other interventions have failed to produce weight loss. Given the great success, dramatic health improvements and lack of alternative options, it perhaps seems strange that we are not doing more.

SHOULD WE SLICE OUR WAY TO THINNESS?

As with any surgery, the reality of bariatric operations is that there are a number of risks. It is estimated that 0.5 per cent of procedures result in death. Although this might sound a relatively small number, it means that one in 200 people undergoing bariatric surgery will die during the operation or shortly afterwards. Given that fenfluramine was withdrawn after causing heart defects in one in 17,000 people, the surgical risks seem significant. It is true that in some of the most experienced

* 485, 551 and 661 pounds respectively.

and professional clinics the risks of surgery are considerably lower, but it remains a serious and dramatic procedure, and not to be undertaken lightly.

There are also many potential long-term complications of having your insides altered in such a dramatic way. Common side effects include gallstones, hernias, malnutrition, stomach perforation, ulcers, vomiting, diarrhoea, low blood sugar and bowel obstructions. Patients have been known to starve despite access to food, unable to get enough nutrition into their bodies. And the rapid post-operation weight loss often means that further surgery is needed to remove skin folds, with all the complications and risks that this brings.

Surgical procedures are evaluated in a different way to drug treatments, and judged by alternative standards. If a widely prescribed drug killed one in 200 people and caused such dramatic long-term difficulties, there would be an international scandal, and multibillion-dollar lawsuits. But for surgery, those risks are considered acceptable. And this is particularly the case when it comes to the only really successful treatment for obesity. Fiona Willer is an Australian dietitian and researcher, and vocal critic of surgical approaches to weight loss. She told me:

> The risks are incredibly high, considering that there's nothing actually wrong with the organ they're operating on to start with. If it was a plane, you wouldn't fly in it. Slicing up the guts of two thirds of the adult population is clearly not feasible, yet surgeons get put up on this pedestal as heroes who are going to 'fix the obesity epidemic'. It is reprehensible the amount of free rein bariatric surgeons get in the media. There is a cultural acceptance that fat people should be willing to go through experimental procedures. Patients are being used as lab rats, being made to pay for it in a lot of cases, and other than ending up malnourished if they don't closely follow their eating plan, we don't really know the consequences in the long term yet.

The psychologist Dr Deb Burgard told me:

> Weight stigma provides a fertile ground for the idea that body size is a disease. Many medical people argue that this is progress, that it is better for people to view higher weight as an illness than a

sin. But gay people in the 1950s were not served by medical 'interventions' to lobotomise or shock them or institutionalise them. Things get better when body characteristics are considered neither sin nor illness but rather a normal human expression of diversity. Higher weight people have not yet found their voice and power to rise up in huge numbers and say no to surgeries that mutilate our healthy organs and starvation regimens that result in worse health, but eventually I believe we will.

There are also enormous psychological implications that can result from such rapid and dramatic changes to the body. Professor of Eating Disorders Ulrike Schmidt told me:

> You really need to look at any underlying problems in preparation for surgery. Is their weight gain psychologically driven? Are they just lonely and sad, with nothing else in their life, because if they are, then surgery won't resolve it. Some people are given surgery and cannot cope. They will continue to binge, often overeating on semi-liquids like ice cream. Good surgical programmes address the reasons for someone getting to a point where they need surgery.

It should be remembered that 50 to 60 per cent of bariatric surgery patients are thought to suffer from binge eating disorder, a condition often linked to severe trauma. Although this sort of eating might be destructive, it is also a coping mechanism. To force people's bodies to stop doing it might well leave a void that needs filling. Nichola Ludlam-Raine again:

> Often there are serious underlying issues, including child or sexual abuse. You have no idea what that person has been through. Anxiety, post-traumatic stress disorder and depression are often reported in severely obese patients. Food is sometimes the only way they have of pacifying emotions, so you need to be careful when you take that away. People can really suffer psychologically. They can have problems with friends and relationships. Sometimes when a 'fat friend' becomes a thin friend, people don't like it, and you need to make sure they are prepared for that.

Long term, the results of bariatric surgery are more mixed. Many patients end up dilating their remaining stomach and returning to old ways of eating. Surgery removes the physical desire for food, but given that severe obesity is so often caused by deeper psychological needs, it does not work for everyone.

Sven Schubert saw more mixed results after the initial success of his surgery, something he puts down to not being fully prepared for the changes. 'After the initial loss, my weight started to creep up. Regain is expected, but it seemed more of a disappointment to me than the medical community. The new feeling of being full is really positive, but you suddenly need to create new coping strategies for stress and upset. You really need help to turn your life around before you get the tools.'

Without adequate support through surgery, many turn to drugs and alcohol, a surrogate for the comfort that food can no longer provide.[10] [11] While surgery may well be lifesaving and the benefits can still outweigh the risks in the most seriously obese patients,[12] bariatric surgery should remain a last resort. Although there have been many calls for the UK to perform more surgeries to help combat obesity, often based on cost effectiveness,[13] unless we are confident we know all the long-term risks involved, we should take care to prevent damaging people further. Such dramatic, life-changing surgery requires psychological support every step of the way. If we cannot provide that, we should really be looking at other ways of helping people. Thinner is not always healthier, and the cheapest treatments are not always the best.

WHAT CAN IT TEACH US?

It is a little depressing that in an era of astonishing scientific progress and medical breakthroughs, when it comes to obesity, a brutal surgery dating back to the 1950s is the best option that we have. But while bariatric surgery will doubtlessly continue to spark intense debate, it does raise a question that is perhaps even more interesting. How exactly does it work? Surely if we can discover that, we can create new and revolutionary drug treatments. If the changes surgery causes to appetite, blood sugar and set-point weight could be replicated in a simple pill, without chopping out intestinal piping, then this would open the way for truly game-changing obesity and diabetes treatments. Almost all the risks of bariatric surgery are due to its invasive and permanent nature. Take those away, and obesity, insulin resistance and even type 2 diabetes might become easily treatable conditions. Perhaps even things of the past.

Unsurprisingly, this has become a hot area of research, with many candidate hormones showing potentially promising effects. It is, however, fiercely complex, and many of the interactions are barely understood. Our guts produce a myriad of different hormones, and as it is extremely hard to perform equivalent surgeries in laboratory animals, most of the information we currently have comes from observing human subjects.

A few of the hormones involved, particularly GLP-1, peptide YY, pancreatic polypeptide and oxyntomodulin have been identified and have shown some promise, leading to the hope that, one day, the benefits of bariatric surgery might come in a simple, safer and more acceptable form.[14] Although this is possible, as with anything else that interferes with hunger, the likelihood is that there will be complications and difficulties. And as these will be drug treatments and not surgeries, they will have to be considerably safer than the operations they replace. Presuming such treatments are a long way off, and that there is nothing else around the corner, there is a huge void. There are currently no effective drugs. Surgery might be effective but it is also brutal and risky, and only really acceptable when people's lives are in danger.

So what to do? If you can't diet, drug or cut your way to thinness, where should you turn?

There are many people out there who think that a whole new approach is needed to the treatment of obesity. With most other avenues failing to deliver results, these voices are starting to make an increasing amount of sense.

18
NON-DIETING

*Loving yourself is not antithetical to health, it is intrinsic to health.
You can't take good care of a thing you hate.*
Lindy West, *Shrill: Notes from a Loud Woman*

It can be an emotional thing when you challenge deeply held beliefs. But with so little success being seen in measures to tackle obesity, a few people out there have started to do exactly that. Some have suggested that the pursuit of weight loss is not only incredibly difficult, it might actually be harmful and counterproductive. So instead of wasting so much time and effort trying to find ways of making everybody thin, perhaps we should just . . . stop. Maybe some people are fat, and we should just leave them to it.

This is of course a controversial approach. There are millions of researchers, medical professionals, diet industry employees, public health workers and health gurus who make their living from people's desire to lose weight. And let's face it, despite a good deal of controversy and misinformation, being extremely overweight is definitely bad for your health. Really bad. It makes people miserable and sick. It is linked to large numbers of deaths from a diverse set of conditions, and costs healthcare services billions of pounds. To give up on fat would be like giving up on smoking prevention back in the 1970s.

For many people, in fact probably for most people, any suggestion of just dropping the issue is inconceivable. The thought of abandoning the focus on weight loss when it comes to overweight or obese individuals raises hackles like nothing else. It attacks the fundamentally held belief that being fat is bad and we should do everything we can to fight it, even knowing it will end in failure.

For much of the world, including most of the scientific community that works on obesity, anyone claiming that we should give up on restrictive weight loss diets is considered a heretic or, worse still, a feminist. Some activists have a history of framing the imposition of diet culture upon women as an act of male aggression, as violence imposed by the patriarchy, forcing women to deny and constrain themselves. The creation of an unattainable thin ideal curtails women's physical power, making them controllable and malleable. Think of the Minnesota starvation experiment volunteers, weakened and broken from lack of food, unable to work, think or organise themselves. Women have been convinced they must diet to be happy, ensuring that they never are.

Out of these beliefs, fat acceptance movements have grown – a branch of radical feminism that rejects all diet culture and encourages women to love and accept themselves as they are. And from this, an anti-diet movement has been created. Although it did not formally enter scientific literature until the early 2000s, 'Health at Every Size' has its roots in the 1970s and '80s, when groups of disparate dissenters, frustrated and angered by the oppressive nature of diet culture, rejected the notion of weight loss as an aesthetic or health ideal. Health at Every Size (HAES) has since been trademarked by the Association for Size Diversity and Health. The movement is not unified, but at its core is the belief that the focus on weight loss is extremely harmful and should be stopped.

Perhaps because of their roots in activism, non-diet movements are often not taken very seriously within academic circles. There is mockery and derision, with many citing the seemingly overwhelming evidence of increased health risks seen in obese patients as reason to dismiss any anti-diet arguments. It is all very well telling fat people to love themselves as they are, but when someone has a BMI of 45, a history of diabetes, high blood pressure and life-threatening cardiovascular problems, self-acceptance isn't going to save them.

The problem with any ideological movement is that it can struggle to remain objective, with a tendency to cherry-pick evidence. But there are increasing numbers of informed and intelligent people adopting some of the principles of the non-diet movements and exploring what they might mean in the modern world. And although the current soaring rates of obesity might cause many to think that fat acceptance should be abandoned, some believe that this makes it more important than ever.

WEIGHT ACCEPTANCE

The current Wikipedia entry for Health at Every Size describes it as a 'pseudoscientific theory' whose 'main tenet involves the rejection of overwhelming evidence and the scientific consensus'. This is perhaps not helped by the 'Health at Every Size' name, something that Deb Burgard, one of the founders, described to me as being 'not ideal'. Many believe that weight-inclusive approaches such as HAES simply encourage people to give up any idea of personal responsibility, rejecting all evidence that living in a larger body harms your health.

In reality, the HAES movement encourages a weight-inclusive approach to healthcare, trying to maximise behaviours that improve physical well-being, without ever focusing on a patient's weight. It rejects the notion that BMI, fat or body size is an accurate proxy for health, but as we have seen already, this is not too far from the truth. They also suggest that a natural diversity of weights exists within any given population, which really is the truth, and believe that trying to make everyone attain an arbitrary definition of normal might be harming some individuals.

Rejecting any notion of controlled or restricted dieting, the movement has spawned an approach to food termed 'Intuitive Eating'. This is based on the idea that people should be encouraged to eat whatever they desire, tuning into their natural hunger signals. Prescribed diets, calorie counts, forbidden items and strict eating rules are abandoned in favour of understanding what your body is telling you, and developing a better relationship with food. Again, this is deeply controversial, undermining many years of healthy eating advice and a multibillion-pound diet industry. Surely if people are allowed to eat whatever they desire, they will just eat Mars bars, pork-pie sandwiches and deep-fried cake?

Perhaps most importantly, the movement encourages a world that does not stigmatise fat people, accepts size diversity, and aims for improvement in healthy behaviours without a laser focus on weight.

Stigma, the movement claims, is the greatest barrier to fat people being healthy, and so any approach that encourages it should be avoided at all costs. As a result, size acceptance movements have been some of the most vocal critics of the 'obesity epidemic' narrative, claiming that it only results in more stigma and harm. This puts them in conflict with many of the world's large public health bodies, which doesn't help their outsider status.

Until quite recently, weight-inclusive practitioners have been a small subset of the healthcare world, mostly dietitians, psychologists or nutritionists offering their services to a limited number of private clients. But in recent years, Health at Every Size, and similar approaches, have started to suggest that they may have something to offer beyond clinical practice. Given the failure of most public health interventions to tackle obesity, perhaps a weight-inclusive approach might benefit this field too, shifting the focus towards improving health, which is eminently possible, rather than reducing size, which is very difficult.[1] Here, things become even more controversial, particularly as the normalisation of body weight is currently at the top of every public health agenda around the world.

Laura Thomas, PhD, is a UK-based registered nutritionist who specialises in Intuitive Eating and non-diet nutrition. She explains that there is great resistance to some of the HAES approaches, partly because of a lack of evidence. She told me:

> There is some evidence that intuitive eating can be effective, especially in terms of improving metrics of mental health and normalising eating behaviours but there are lots of unanswered questions and lots of work to do. We need more robust randomised controlled trials,* but it is a challenge for anyone to get funding because everything goes to the obesity epidemic. I am the first to admit that there is not enough evidence, but the weight focused approach is not evidence based either, and there is a huge problem of weight bias and stigma in obesity research. There is good evidence that diets don't work and might be harmful. A weight inclusive approach is less stigmatising, kinder and more compassionate.

* This is a type of experiment designed to remove bias when testing the effects of an intervention or treatment. It is particularly hard to do with diet.

We have discussed the ineffectiveness of diets already, but Laura mentions side effects that are also worth considering. Studies have shown that restrictive weight loss dieting is associated with binge eating disorder, bulimia, future weight gain, negative body image, low self-esteem and depression, with particularly stark effects among children and teenagers.[2] The psychologist Deb Burgard explains:

> Part of the problem is the underlying assumption that we are all meant to be one size. If we took 100 babies from around the world and could somehow feed them exactly the same way, would we expect them to all reach the same height as adults? No. So why do we expect people to be the same weight? I don't see the scientific basis for considering a body size a disease – it is clear we are not meant to be one size. Some people are like mastiffs, some are like poodles, and some are like greyhounds. If you starve a mastiff, it will never become a poodle.

A focus on losing weight through restrictive dieting, the most commonly prescribed healthcare intervention for obesity, comes with many known dangers. But perhaps these risks are deemed acceptable when compared against the harm caused by being fat? This might be the case, but with the distinct lack of long-term success associated with most dieting, it is worth considering if these dangers are worthwhile.

Dieting and a pervasive fear of obesity can cause problems for many vulnerable people, not least mental health issues. According to Professor Ulrike Schmidt, 20–30 per cent of people with anorexia and 40–50 per cent of people with bulimia will have a history of obesity. That was certainly the experience of Grace Victory, a blogger, YouTube star and size acceptance campaigner. She has struggled in the past with weight issues and eating disorders, but considers that now she has adopted a non-diet approach to her health, her life has dramatically improved. She shared her story with me:

> I remember being aware of my weight from about eight or nine. I was called fat for the first time at twelve and decided I didn't want to eat, or I would throw up food after eating. I just assumed that there was something wrong with the way I was. Between twelve and twenty-one I had a cycle of not caring, gaining

weight, binging, dieting and restriction. I thought that was the normal way to live. For all those years, I thought thin people had the trick and kept to it. It made me very depressed and I had incredibly low self-esteem. I was terrified of being big and thought, 'How can you be happy if you are bigger than the ideal?' I managed to lose weight through Slimming World and Weight Watchers, but I became anxious about a lot of things. I was scared of oil. I was scared of going out. Eventually I just snapped and ate what I wanted, but I felt incredibly guilty afterwards. Not long after that I went into eating disorder treatment. I was never thin enough to be classified as anorexic. I had binge eating disorder with an anorexic mind set. The thing is, when I was younger, people would actually praise me when I was restricting. Now I am comfortable, but only after eighteen months of working on my own mind daily. I just try to eat what my body tells me to, and I am happier and healthier than I have ever been. But people still get offended that I am not a 'good fat person', not exercising, restricting or trying to lose weight. They get offended that I wear nice clothes and appear confident.

By dropping our restrictive, rule-based approach to food, it might also relieve some of the mental stress that comes with it.

STIGMA AND SHAME

At the heart of the weight-inclusive movement is the reduction of stigma and shame. It asserts that any approach demanding everyone achieves a particular weight, and categorising a certain range of BMIs as 'normal', is likely to create stigma. And stigma is known to cause many health inequalities,[3] including those frequently associated with obesity and assumed to be caused by fat. People who are oppressed get sick far more frequently than those who are not, and perhaps much of the conventional policy approaches to obesity contribute to harm in this way.

With stigmatising messages about fat so prevalent in modern society, it is probably impossible to accurately assess how damaging this prejudice might be. Sadly, there are no obese populations not subjected to prejudice, and there is no experimental way of removing its effects.

But studies have shown that the degree to which people are dissatisfied with their weight, defined as the difference between the weight they are and the weight they would like to be, is related to their risk of developing type 2 diabetes, independent of their actual BMI.[4] This raises the possibility that the stress caused by telling people they should be thinner than they are is one of the things making them sick in the first place. Sven Schubert, the patient advocate, told me: 'Stigma is crushing people. The effort to overcome your body is hard enough, without it being stigmatised. The world tells you to stop eating and start moving, which is the worst advice. It's not like you haven't thought of that. It's like telling people to slow down and stop breathing, it just makes panic set in.' Deb Burgard agrees.

> Many people who get classified as 'obese' will be mastiffs. If you burden them with interventions that are supposed to change them into poodles, you are telling them from the start that there is something wrong with being a mastiff. That is inherently stigmatising. There is a whole industry that depends on people's fears of not being thin enough, white enough, healthy enough. Health and weight have become a politically palatable basis for carrying out racism, as if people who have more health or thinness have more worth. And stigma may be a great way to get people to try a practice that they think will get them out of the stigmatised group, but it is a terrible motivation for developing a sustained practice to support your well-being.

If anyone can be said to have kicked off the rise of fat acceptance movements, it is the British psychologist Susie Orbach. Her book *Fat is a Feminist Issue* was published in 1978, and she has spent many years campaigning against the pressures on women and girls to control and restrict their bodies. She currently works in clinical practice, helping people with food and body-image issues. She told me:

> We no longer teach kids to eat when they are hungry and we inadvertently talk about food as quasi magical. This creates conditions for disorder and anxiety. We have a huge prejudice around size, and people are constantly encouraged to think that their body is if not the most important thing about them, then

the thing they need to get right. We have six year olds playing cosmetic surgery games now.

Interestingly, she also believes that society has moved on since her highly influential book was published, though not in the right direction. Whereas body-image pressures were once a solely female domain, these days the net has been widened.

> I used to say there were differences between men and women. It has always been very profitable to destabilise women's bodies, but it has evolved to include men's now. There are large industries whose business it is to increase these problems. The food industry, the fashion industry, the beauty industry, the cosmetic surgery industry, the diet industry – the latter who should be prosecuted for false advertising.

Another consequence of the stigma around obesity and the focus on weight loss approaches has been to create a state of learned helplessness among many fat people. For anyone with a BMI of 50, particularly if they have experienced a lifetime of failed dieting, the thin ideal seems a distant and impossible dream. When people don't see the point in trying, they give up any idea of working towards their own benefit. A weight-inclusive approach may help to counter this problem, encouraging eating well for pleasure, exercise for enjoyment, and adopting healthy behaviours for their own sake, not in the pursuit of some unachievable goal. It meets people closer to where they are, and so leads to sustainable change.

An understanding of learned helplessness has a history of success in the battle against the AIDS epidemic. Early information campaigns focused on shock and fear, doing an effective job of raising awareness. But in many marginalised groups, particularly intravenous drug users and young homosexual men, the behaviour changes required were beyond what was easily achievable, and some of the messages deeply stigmatising. As a result, many developed a certain fatalism about the disease. It was not until charities and public health campaigns started to understand and engage with these groups, meeting them where they were rather than setting unattainable goals, that things started to improve.

Although deeply controversial at the time among conservatives, needle exchanges, safe sex information and free condoms saved the lives of countless people, and helped contain the spread of this vile disease.⁵ It is interesting that those who opposed these measures believed they would only encourage and endorse morally questionable behaviours – much the same language is used to pressure obese people today.

INTUITIVE EATING

Of course, abandoning dieting might lead people to eat badly and damage their health. With this in mind, in the mid-1990s, dietitians Evelyn Tribole and Elyse Resch developed an approach to food called 'Intuitive Eating'. This was driven by their belief that most people, particularly those who have a history of restricted dieting, have lost touch with their body's sensations. Intuitive Eating encourages people to reconnect with these signals, and trust what their body tells them when deciding what to eat. US dietitian Christy Harrison teaches these techniques. She explained to me that 'when people constantly repress hunger cues they become escalated and fullness cues are played down. Over time, people can be taught to sense hunger more, and pay attention to it. They can learn what their body is telling them.' This approach is controversial, largely because it makes no promises about reducing people's body fat. Many come to it with this goal in mind, and might find themselves disappointed. The way Christy Harrison sees it, 'Intuitive eating is not a weight loss approach. You might actually gain weight, but what is the alternative? Any approach that is about restriction will result in weight cycling, which is damaging to health in its own right, and weight cycling also typically leads to people getting heavier over time. With intuitive eating, people reach a weight and stay there.'

The endocrinology professor Gareth Leng explained to me why restrictive dieting might have a relationship to future weight gain, perhaps suggesting that a more intuitive approach might be preferable in the long run:

> Controlled eating is a predictor of weight gain, but it is hard to know if it is cause or effect. Controlled eaters disregard signals from the body and could undermine control mechanisms. So encouraging controlled eating is perhaps the wrong thing to do.

Perhaps in an obese state when stuff is broken, controlled eating might be the only way to bring things back, but it is probably not good for the whole population.

There is increasing evidence that Intuitive Eating might benefit some individuals. For anyone who has been worrying about the retired boxers and weightlifters we discussed in Part 1, Intuitive Eating has been shown to help them re-engage with their bodies' natural signals and avoid some of their post-career problems.[6] In many others it has been shown to help improve health, perhaps because it reduces the level of conflict between them and their bodies. Forever battling with your size is a horrible way to live, and for many, releasing that tension might cause improvement.

But as Gareth Leng suggests, the approach is unlikely to suit everyone. For some, natural satiety signals might not be functioning properly for any number of reasons, including endocrine dysfunction. Others might be genetically inclined to have an unhealthy body size or have satiety signals so badly disrupted that they can never fully reconnect with them. Even Intuitive Eating's strongest advocates admit that it might not be for all. Laura Thomas told me: 'It is not that weight is not important. You work differently with someone with a BMI of 18 than you do with a BMI of 35. It is very reductionist to assume that one approach will work for everyone. There are all types of reasons for someone not responding to hunger and satiety. There are lots of ways of working with someone and there will be lots of caveats.'

The Intuitive Eating approach might give us clues as to why some people seem more susceptible to the effects of an obesogenic environment. It might also explain why in countries where controlled dieting is hugely prevalent, resistance to that environment seems to be diminished. Christy Harrison told me: 'Once people have permission to eat things, they lose their lustre. People eventually allow themselves more access and they lose the idea of forbidden fruits. When we think about taking pleasure in things, we actually find that people don't feel compelled to eat as much in order to feel satisfied – and often those formerly forbidden foods are not as nice as people think they are.' On this Deb Burgard told me:

> This argument that the environment is 'obesogenic' implies that food is irresistible and we cannot control ourselves. But where

is that experience coming from? People who are restricting and people who have food insecurity are much more vulnerable to this experience of food being so salient. When your weight is not suppressed and when you have access to nourishing and good tasting food, then food is much more likely to just be food, and a person does not need to make eating decisions based on the temporary availability of it. So we need to look at who has an interest in people having food insecurity – from financial need or access or intentional restriction. Look at how many corporations are both food and diet companies.

If Intuitive Eating achieves anything, it is to encourage people to have more resistance to the world around them. For when life is about restriction and control, a world that offers a Twix or a Snickers at every turn is hard to resist. You can fight hunger for long periods of time, but you only have to give in to it once in a while. But when given control, people rarely eat chocolate all the time. I spent two years working as a pastry chef, with enviable access to the finest chocolate, cream, butter, caramel, cakes, fruits and desserts. But other than essential tasting, it was rare that I ate much sweet food, and spent far more of my time bartering with other kitchen sections for chicken, fish and vegetables. The denial of diet culture, and the pervasive guilt that so many people feel, makes environments obesogenic. Dieting fosters a belief that forbidden and bad foods must be uniquely delicious, even though they rarely are. When forbidden fruit is no longer forbidden, it's just . . . fruit.

WHEN DIETS WORK

As a chef who loves to eat I am naturally inclined towards these ideas, especially as I strongly feel that individualised diets draw people away from food's true meaning, something we shall return to before the end of this book. But even I have to admit that for many people, restrictive dieting is an important strategy in maintaining good health. There are a million good reasons why someone might want to shift some pounds, and it is not my place to tell them what they should do. For all the bad science and ill-conceived fads that I like to mock, there is a need for something that stands a good chance of working.

Nichola Ludlam-Raine's work as a dietitian brings her into contact with severely obese patients, and she runs weight loss groups to help them develop strategies that might improve their health. Her clinics have seen many successes over the years and drawn praise from health authorities and academics across the country. Although many people view dietitians as nagging or judgemental, that is far from the reality of Nichola's work. She explained her approach:

> I run twelve-week weight loss groups, and once finished, I offer individuals four follow up sessions a year. If I can help people to lose just five per cent of their body weight, then often they will feel amazing. Even those who don't lose weight can see improvements in their blood work and overall sense of well-being. The programmes are not all about diet. I really try to take things back to basics. We work on understanding calories and energy density. We look at coping strategies for emotional eating and helping people change habits. We run 'supermarket tours' to help people shop in a better way. It is about education and baby steps. There is a lot of work on different strategies to create behavioural change. I don't create meal plans; instead I help people to create their own. Success happens when you are a facilitator, not an educator.

Dr Yoni Freedhoff is the founder of the Bariatric Medical Institute – a multidisciplinary weight management centre in Ottawa that focuses on non-surgical treatments. He is a medical doctor, who moved to weight management after being repeatedly approached by patients asking for help in this area. He told me:

> There is not one diet, not one strategy. We look at patients' social lives, diet history and help them find the healthiest life they can enjoy. Life is complex and if you just say 'eat less, move more' the effect falls away pretty soon. Obesity is a chronic condition, not a temporary one, so we help people to change their behaviours to improve their quality of life.

On the success of dieting, he says:

> Weight or BMI based goalposts for success aren't useful, and may well be futile, if they require total losses. But you can help people

achieve a 5 to 10 per cent weight loss. The Look AHEAD study showed that 35 per cent of people maintain five per cent, which is far from 'the only people who maintain weight losses are unicorns'. The outcomes we look for are not the same as the commercial weight loss industry, which is completely toxic. It is also not the same as much of the medical community. If a clinician says to someone with a BMI of 45 that they need to get below 25, then I am not really sure what the point of that conversation was.

Similarly, Nichola Ludlam-Raine says: 'It is far easier working with people who have never tried to diet or lose weight before; for example a middle-aged man who has slowly put on weight over the years, and then found himself needing to lose it. When you work with long-term yo-yo dieters, they tend to have preconceived ideas and think, "I've tried this before, why will it be any different this time?"'

In intensive, patient-focused approaches, there is little doubt that some weight loss can be achieved and certainly no doubt at all that health improvements can result. But even for the most well-supported scientific studies, large amounts of weight loss seem hard to attain.

Participants in the US Diabetes Prevention Program (DPP), a four-year experiment to study the effects of lifestyle interventions on the risk of developing type 2 diabetes, received regular consultations with individual lifestyle coaches, intensive training in behavioural management strategies, supervised physical activity and extensive advice on dietary modification. All the help received was tailored to individual needs and modified if it was not working. Participants achieved an average weight loss of 7 kilograms (15 pounds) in the first year, but in long-term follow ups, most of this weight had been regained.[7]

The Look AHEAD study that Dr Freedhoff mentioned was a ten-year programme that went even further, offering participants regular access to teams of lifestyle coaches, dietitians, psychologists, doctors and physiologists. For the first year everyone attended three to four sessions a month, a combination of group and individual activities, as well as regular email and telephone contacts. The tools used were altered to match individual needs, using the best techniques to achieve maximum results. But even then, over the long term, only a third of people were able to maintain a weight loss of 5 per cent. After ten years of very expensive guidance and hand holding, using the very best methods available, people only lost a few pounds.

It is well worth noting that although the Diabetes Prevention Programme and the Look AHEAD study both failed when it comes to the sort of significant long-term weight loss most dieters desire, both did show marked improvements to participants' health. DPP focused on diabetes, pitching lifestyle against a well-known drug therapy, and proved that the lifestyle interventions were far superior. Exercise particularly seemed to have a significant effect, even if initial weight loss was not sustained at all. The Look AHEAD study may have had modest weight loss results, and was stopped early because it failed to show enough impact on cardiovascular risk, but participants showed improvements in blood pressure, sleep apnoea, visceral fat, depression, kidney problems, physical mobility, the need for diabetes medicines, life quality, knee pain, sexual function and inflammation.[8] So it is not always wise to assess dietary success by weight loss alone.

WEIGHT OR BEHAVIOUR

In truth, although they often seem miles apart, there is a great deal of common ground between the messages of movements such as Health at Every Size and more conventional weight management. Both value and encourage exercise. Both push for a varied, balanced diet and good quality sleep. Both want people to be as healthy as they can be. It is just that conventional approaches ask people to do these things with the goal of losing weight in mind. If it is the behaviour change that we are looking for, the most important thing is to find an approach that encourages sustainable shifts.

When smoking, excessive alcohol consumption, low fruit and vegetable intake and a lack of exercise are studied, it is no surprise that people doing all those things have high rates of mortality. Combine that with being fat, and the picture looks bad indeed. Anyone fat who eats badly, drinks too much, smokes and rarely gets off the couch is far more likely to die than a thin person doing exactly the same. But curiously, research has shown that as these behaviours change, the difference between fat and thin people seems to diminish. In fact, for non-smoking, regular exercisers who only drink moderately and eat lots of vegetables, there is no difference in mortality rates between thin and fat groups.[9] Similarly, when exercise alone is looked at, the risk of developing CVD has been shown to be nearly identical for obese or overweight men with high levels of fitness, with differences along BMI lines only occurring across unfit groups.[10] Maybe it is worth considering that it is not always a lighter destination that improves people's health, but the journey required to get there.

THE HEALTH OF THE PUBLIC

Intuitive Eating and Health at Every Size are controversial approaches, and aren't for everyone. But is there anything they can teach the world of public health? Certainly there is a lot that public campaigns could learn when it comes to stigmatisation of obese people, and there has been some progress in this area. It can also be argued that a focus on behaviours rather than weight is preferable. Obesity is not an activity that people do, and so the consistent demand to stop being obese is of little value. Studies have shown that the public are far more likely to be engaged with campaigns that highlight behavioural improvements, rather than those that spread fear and disgust.[11]

Most of all, the existence of campaigners for a weight-inclusive approach should help us remember a fundamental truth about public health organisations. They exist to help us be healthy, not to become thin. Anyone who researches, campaigns, writes or works on obesity, food and health should ask themselves this. Are we worried about people's health, or about their silhouette? If we could magically take away the health harms of being fat, would we consider the problem solved? Or are we just desperate to make people thin to satisfy aesthetic and cultural ideals? Because if the fight is really about thinness, then we need to ask ourselves some serious questions.

Many people have very little choice about how much they weigh. Equally, some people do not care one jot if they are fat, and are not interested in losing weight. A fair society will fight to keep all these people as healthy as possible, not demand the impossible of them. However much we want people to be thin, some will remain the mastiffs that they are, and never contort themselves into greyhounds. So while they have fat bodies, care must be taken not to exclude them from paths to better health. If they can be encouraged to change a few things about their lives, whether it is diet, sleep, exercise, mental well-being or substance use, their health will be improved.

At its best, Health at Every Size is not against weight loss per se, but it does challenge the single-minded pursuit of weight loss. As we have seen throughout this book, losing weight is really difficult, and for many it will be impossible to achieve in a safe, healthy or humane way. So with this in mind, an approach that looks to improve those people's health regardless of weight is far more inclusive and potentially more effective. Where HAES can fall down is when it steps into a blanket condemnation of all dieting, and a rejection of anyone who takes that path. There will never be a successful one-size-fits-all approach when it comes to health. Some of the most vulnerable and at risk people might need helping through weight loss in a safe, sustainable way. For others, the only alternatives to dietary restriction are surgery or death. And considerable numbers of people want to lose weight for aesthetic reasons, and should not be judged either.

Grace Victory receives countless calls for help and advice through her social media, often from young girls struggling with complex food and body image issues. 'I try to make advice individual,' she says. 'Body image is hugely complex and everyone is different. But I do tell people that we have been brainwashed. We have been told that you have to be a certain size and colour to be beautiful. I try to tell people that we are far more than what we look like, that identity is more than just your size.'

It is a shame that there is so much division between the weight-inclusive movement and more conventional approaches, because they share a good deal of common ground. But at the public level, there is still a dominant public health agenda claiming that everyone submit to the same weight goal, and it is hard not to conclude that this is damaging people. Accepting the existence of body diversity and spreading more inclusive messages about positive behaviour change would surely be constructive. Sven Schubert told me: 'If someone had taken me by the hand and said this is why you are different and this is what you can do, it would have been hugely liberating. But I was always 100 per cent convinced it was all my fault. Once I had the knowledge that there was a genetic component, and that my body responds differently, it took the weight off my shoulders. It meant that I could look for help.'

Healthcare should be about including people, caring for every aspect of their well-being, and giving them the tools to live the best life that they can. The moment it strays from that mission, it has gone beyond its remit. Although Health at Every Size sometimes takes an extreme line, it is a movement that encourages us to care for everyone, considering every aspect of health. And that is something we can learn from.

19
HOW THE FRENCH LUNCH

*I sometimes worry that by encouraging so many more people to try
their hand at baking . . . I'm going to find myself in court one day
charged with accelerating the national epidemic of obesity. To
which I will plead not guilty. A slice of Victoria sandwich is never
going to harm anyone.*

Paul Hollywood

There is a factor that is thought to contribute significantly to obesity,
conveying an increased risk that is far greater than the well-known
and much discussed socio-economic factors.[1] It is also something
with the potential to undergo rapid change at population level, and
could theoretically be a target for interventions that are cheaper and
more achievable than trying to lift everyone out of poverty. Yet it is
rarely mentioned in debates and never the focus of public health
messages. Indeed, no one seems to consider it an important historical
factor at all.

This factor is our personality. When people's personalities are
assessed by psychologists, creating assessments of the so-called Big
Five traits, a number of strong correlations to obesity are seen. The five
traits assessed, for which people can be high or low, are Openness,
Conscientiousness, Extraversion, Agreeableness and Neuroticism,
creating the suspiciously convenient acronym OCEAN. For some of
these factors, correlations with body weight are weak and inconsistent.
Openness, extraversion and agreeableness have been linked to obesity
in the past, but the relationships differ across groups, nationalities and
sexes, often depending on the method of assessment. But the other

two, neuroticism and conscientiousness, show solid and consistent links to obesity.

With personality traits, there is of course the possibility of a two-way relationship when it comes to obesity. As we have seen, the experience of living in an obese body can be traumatic and difficult, and liable to alter someone's personality. But studies have shown that although personality can change over a lifetime, the likelihood and direction of that change is not influenced by BMI.[2] It seems that certain types of personality can greatly increase your chances of becoming fat, but being fat is unlikely to change your personality.

THE UNSURPRISING BENEFITS OF BEING A BIT DULL

Perhaps unsurprisingly, highly conscientious people seem less likely to gain weight over their lifetime. The main traits of conscientiousness are being vigilant, organised and self-disciplined. They tend to engage well with public health messages, regularly undertake positive behaviours and worry about their weight a good deal.

It will be no shock to learn that this trait is strongly correlated to other good health outcomes.[3] Conscientious people are more likely to visit the doctor regularly, check smoke alarms, exercise and eat a good-quality diet. They are far less likely to drink heavily, use drugs, drive quickly or smoke cigarettes. To be honest, however, despite their enviably long lives, they do not sound that much fun to be around. Even when they don't live to 110, it probably seems about that long.

That said, if you really want to live a long life, scoring high for conscientiousness is a good place to start. Conscientious people are likely to maximise their chances of never becoming overweight, and even if they do get fat, their personality and behaviour will reduce their odds of getting sick.

But alas, we are not all conscientious, and the world is probably a more interesting place because of it. It also seems unlikely that this trait has diminished much over the past fifty years, especially when we look at the recent decline in behaviours such as smoking. So it is perhaps more interesting to consider the second personality trait consistently linked to obesity, and that is neuroticism. Because here things get a good deal more interesting.

Neuroticism is a trait that is broadly described as the extent to which people are susceptible to stress. This can take many forms, including anxiety, aggressive behaviour and depression. A high level of neuroticism indicates a low level of emotional stability, and will generally be associated with poor impulse control. Highly neurotic people have an increased risk of obesity, but interestingly, neuroticism is also strongly related to the chances of developing an eating disorder.

Neuroticism is associated with being at either extreme of the BMI spectrum, either underweight or obese, and with all manner of disordered food behaviours.[4] Highly neurotic people tend to be deeply concerned with body image, health and weight, but with a tendency to lose control. And even when they are of normal BMI, they are at risk of a number of health problems, including elevated blood triglycerides, hypertension and inflammation. To make matters worse, they tend to have higher cortisol responses to external stress, something likely to lead to weight gain, a drive for energy-dense foods, and excessive depositing of visceral fat.

Associations between high neuroticism and poor health are consistent across diverse population groups. It is an extremely challenging state for people to live in, and seems to be well correlated to many of the so-called diseases of affluence that affect modern life. There are known mechanisms, most notably the tendency of an increased stress response to cause inflammation, that indicate neuroticism might contribute to the development of these diseases. It all points to this personality trait being a key driver of ill health and obesity.

The question is, has it increased in recent years? Is it possible that a rise in neuroticism has contributed to the growth of obesity, and if so, why? What external factors might have driven us to become more neurotic, anxious and depressed? And what might we do about it?

LOCUS OF CONTROL

Western societies tend to have a high 'locus of control', meaning that people try to take maximum responsibility for their own decisions. Although this sounds like it should be positive for public health, in reality it is anything but. Neuroscientist Dean Burnett explained it to me:

The locus of control is the sense of how much control you have over your own life. In Western societies it tends to be internal, with high levels of individualism. This can turn people into control freaks, meaning they will want to have complete control over what they weigh. But when it is tied to capitalism, there is a sense of entitlement, a feeling that people should be able to have whatever they want, whenever they want it. On the other side, there is an external locus of control, which you see more often in Indian, Chinese and Japanese cultures. With an external locus, you believe you are at the mercy of events, but you also tend to be more capable of self-discipline. Often food is strongly regulated by culture and rituals. Strong community pressures can create motivation to control behaviour. For instance, during Ramadan, Muslims will not eat between sun up and sun down. The brain really picks up on the social aspect and can offer a great deal of control. But if you have a more internal locus, when you see a chocolate bar, you will probably eat it.

Claude Fischler is a social scientist and research director of France's National Centre for Scientific Research. He has studied how commensality, the act of eating together, affects public health. He believes that this external regulation of food consumption has deep roots in our evolutionary past. He told me:

Eating has been a social activity throughout human history. Hunter-gatherers had scarce resources and so had rules to regulate how much people ate. Human societies have always revolved around the regulation of food. But now society is becoming more and more based on the notion of autonomy. It is becoming individualised. We are all developing this Anglo-Saxon cult of individual responsibility. The big difference is that there is now more public health intervention. There is this idea of individual responsibility for everything, and that Public Health will tell you what to do. But more public health regulation is associated with increased rates of obesity.

That might sound counterintuitive, but the number of words in the US dietary guidelines has been shown to track extremely well with rates of obesity. In 1980, the guidelines had fewer than twenty pages. By 2010, that number had risen steadily to nearly 100. Although this might be related to the increased need for action as the health burden of obesity rose, it may also lead us to question whether or not more information is always better.

Psychologist Kimberley Wilson believes that our increased individualism might have further costs, with the potential to contribute to people's weight. She told me:

> As we [society] become more isolated and narcissistic, we develop this idea of individualised success and a winner's mindset. It becomes much harder to have any vulnerabilities. People don't recognise emotional distress, and try to self-soothe through their food and their bodies. Rates of childhood sexual abuse are perhaps much higher than we like to think about. There is lots of evidence that trauma, especially physical trauma, can lead to obesity. People find a way to soothe themselves with food, perhaps because of early experiences of mother's milk bringing calm. It is also thought that some children after trauma internalise the responsibility of attractiveness, and that some might become anti-thin to make themselves unattractive, creating a barrier to protect them from the world.

Perhaps in our individualised society, with lack of social cohesion and a breakdown of anxiety buffers, these problems have intensified, leading many who have experienced trauma to turn in on themselves. Whereas once people lived in protective groups, when we are forced to cope alone we can easily become anxious.

BLUE ZONES AND THE FRENCH PARADOX

It is clear that in many ways, humans have struggled to adjust to the modern world, with its technological progress and social enlightenment. Although for the most part, progress has been immensely positive, we need to change the way we think about food. To create a world where eating is no longer a source of guilt and anxiety, but something that brings us together and helps strengthen our most important bonds. Claude Fischler explains:

We did a comparative survey across six countries and 7000 people, with focus groups and interviews, asking people how they thought about food and eating. The most striking differences were between the US and France. In the US we would ask a woman what she ate for lunch and she would say she was busy and picked up a sandwich. She would tell us exactly how many calories it had, how much fat and carbohydrates. In France, a woman would say 'I didn't eat at lunchtime'. When we asked her some more questions, she would say she was busy and just picked up a sandwich from a bakery. But for her, that was not eating because she ate it alone, not with colleagues or family. For her, consuming something alone was not considered eating.

What do Americans say about eating well? They say you need to make the right, healthy choice. This is driven by Protestant ethics. We know what we should do, but we have guilt and trade-offs, which become problematic. In France, eating well means 'if it tastes good, it is good'. It means eating within the family. It means eating a structured meal and not a snack. In France, we spend far more time eating. Over 50 per cent of the population is eating at 1 p.m. In the UK the peak eating time is 1.15 p.m., but only 17 per cent of the population is eating then.

Since the late 1980s, many epidemiologists have tried to explain the so-called French paradox,[5] the observation that a country with a diet full of rich, fatty foods has strikingly low rates of cardiovascular disease and obesity. Many have insisted that this invalidates conventional nutrition beliefs about fat, but perhaps the real reason has more to do with the meaning of food in that country. Could it be that eating together holds the key, not only to our capacity for restraint, but our resistance to the stresses of the modern world?

Studies of the 'Blue Zones', small pockets of the world with high rates of good health and longevity, show a number of fascinating insights into what factors keep us well. For many years, epidemiologists have picked out small communities where people are likely to be long lived and maintain good health for most of their years. Although modern medicine and public health measures have an impact, Blue Zones are rarely in the most prosperous regions. A group of small villages in the mountains of Sardinia have some of the longest lived men in the world, and yet many are shepherds who spend their lives in relative poverty. The islands of Okinawa in Japan have an extraordinary number of female centenarians, yet most live a humble existence, and experienced desperate hardship in their early years. The Nicoya peninsula of Costa Rica has some of the healthiest elderly people in the world, and yet most live in remote villages with few of the advantages of modernity. And a population of Seventh-Day Adventists in California lead a humble and restrained existence, and have enormously high life expectancies as a result.

Unsurprisingly, many have looked to the diets of these communities. Blue Zone cookbooks have been written, to help people recreate the corn tortillas of Nicoya, the miso soups of Okinawa and the fermented dairy produce of the Sardinian mountains. Yet other than a high consumption of fruits and vegetables, there is very little commonality between the diets of these populations. Most Sardinians enjoy delicious sheep's milk cheeses, washed down with heroically large quantities of red wine. The most popular dish in Okinawa is a stir-fry made from Spam, and the island has Japan's highest concentration of fast food restaurants per capita. Many Seventh-Day Adventists are vegetarian, whereas in Costa Rica, pork and lard are both highly prized.

What unites the Blue Zones more than anything is that people's lives are filled with meaning and purpose. Many have strong spiritual beliefs, and all value the importance of community. The elderly are revered within their families and their long lives are celebrated by all around them. People maintain a useful and valued existence well into old age, caring for grandchildren and often acting as advisers or mentors to their villages. Whether it is through religion, spirituality or family, all their lives are buffered from a fear of death, with the extremely old achieving a state of tranquillity and acceptance.

Although there might not be that much to learn by studying *what* they eat in the Blue Zones, *how* people eat is of great significance. Food is given meaning through sharing, enjoyment and rituals. Rich, indulgent foods are eaten, but usually as part of celebrations or events. Sweet treats are enjoyed and shared with children, without the guilt and recrimination that accompany them in Western countries. And perhaps most importantly, nobody ever diets.

Every Blue Zone has an external locus of control. Even in California, the beating heart of surface vanity and materialism, a small group of Seventh-Day Adventists find meaning in moderation.

It is hardly surprising that the longest lived people exist in strong communities, as our social relationships have a huge effect on our health and longevity. The effect of social isolation is comparable with that of smoking, and far greater than the influence of obesity or a lack of exercise. Being left alone is so devastating that young children in orphanages who have suffered extremes of social isolation have been known to die as a result.[6] And yet loneliness and the breakdown of communities are rarely taken seriously, as we obsessively analyse people's diets, lifestyle choices and activity levels. We are being crushed by a lack of connection and meaning, and yet the world laughs off these concerns as trivial, before insisting people go on a diet.

FINDING MEANING

I am certainly not going to start recommending religion or spirituality as a path to better health, but we can still learn lessons from the benefits that they bring. A world that focuses so much on the individual is not one that our brains evolved to exist in. We are social creatures, and when we turn in on ourselves, we often seem to lose our way. Although there are many things that we could be doing to fix this, I am a chef, so perhaps the one thing I can credibly suggest is that we try to relearn the true meaning of food. Eating is supposed to shape our bonds and relationships, give colour to precious memories, and bring us closer to those we love. It is time that we reconnected with it.

It is not the food that is significant, but the way that we use it. So when you share ice cream with your children, biscuits with a loved one, pizza with friends, or a quinoa salad with colleagues at work, the greatest benefit will not be from the macro- or micronutrients. It will be from the connection and the memories that these moments create. Food has miraculous powers to help and heal, but the notion of longevity diets or life-changing superfoods is absurd when stripped of the context of togetherness and joy.

We look at our modern Western food culture and assume that the nutrient profiles are responsible for any harm. Yet in obsessing about these things, we miss the true problems our modern world creates. We fail to understand what we have lost by changing the meaning of food in our lives. Western diets do not cause harm because of fats, carbs, proteins or chemicals. They cause harm because we eat alone.[7] If we exercise, it's on a treadmill with our earphones in. We eat into our sleep staring into the white light of our smartphones.

This is why I hate diets so much. When we diet, eating becomes a solitary experience. We shut ourselves away and moralise about the choices of others. Dieting makes us believe that health is our sole responsibility. We shun restaurants, birthdays, parties and social events because we will not be able to follow the rules. We distance ourselves from those who share forbidden foods. Hours, days, weeks and lifetimes are wasted dieting, whittling away at our bodies in an attempt to attain an imagined perfection. And our only reward for it is a feeling of failure, a dent in our self-esteem, and the attendant loneliness and shame it brings.

Dieting is the wet dream of neoliberalism, perpetuating the myth that we are responsible for all that befalls us. The fact that dieting does not work, and makes people sick and unhappy, just goes to show how misguided that belief is.

We have been broken by diets. Torn apart by shame, stigma and the cult of individual responsibility. Yet however much these things hurt us, we still manage to convince ourselves that it is all our fault. One day, perhaps, we will realise that we are not to blame, and start to build a better life, and doing that might bring us together.

20

GOING DUTCH

It is better to be vaguely right than exactly wrong.

Carveth Read

In 2017, researchers at the University of Birmingham conducted an ambitious study to assess the effectiveness of interventions designed to help tackle childhood obesity. They took 600 pupils across a number of primary schools, and encouraged them to take part in various activities, all with the aim of improving their diet and activity levels.[1]

Pupils were given the opportunity to spend thirty minutes of every day exercising, and were enrolled in a specialist six-week healthy food and activity programme with their local heroes at Aston Villa Football Club. Throughout the year-long study, families were offered regular healthy eating workshops, and provided with information about opportunities to exercise in their local area. All the interventions were designed after extensive research, following focus groups with parents and consultation with some of the leading experts in the field. Care was taken to ensure that everything on offer was realistic, relevant and, above all, effective.

The results? After a full year of carefully targeted and specially designed interventions, there was no measurable improvement in BMI, diet or activity levels when compared against a control group of pupils who had not received any help. Nothing the researchers had done had made even the slightest difference.

With everything I have learned in the course of researching this book, I would expect a programme such as this to have very little effect on pupils' health or weight. But no effect at all? Even with my inherent scepticism, I would have thought that at the very least the cookery classes might have encouraged children to eat a few more vegetables. Or that the opportunity for regular exercise might have had some sort of long-term benefit. The lack of any positive results just underlines how complicated a problem obesity is.

Well, actually it underlines how complex it is.

Although the difference between complex and complicated might seem like little more than semantics, this subtle point is key to understanding why many such interventions fail. And it might just give us a few clues about the best way to tackle obesity in the future.

You probably remember me promising some positivity in this section of the book, and hopefully I have provided a little of that already. But by the end of this chapter, I can promise you something even more surprising. I am going to tell you how we can solve the problem of obesity. I have a solution that is both simple and complex at the same time. And although a lot of people will not like or agree with it, I can absolutely guarantee that it will work. I bet that's something you weren't expecting to find in this book.

But before I get to this grand solution, I need to explain why tackling obesity is complex, whereas flying to the moon is complicated. And why the difference between the two is very important indeed.

COMPLEX OR COMPLICATED

Your mobile phone is complicated. It is doubtful that anyone knows everything about how it works. It would probably take a few cabinets full of books to fully detail all of its functions and technical specifications. But despite it being fiercely complicated, it can be fully described. And with enough time and effort, it is possible to accurately predict how it will function under any given circumstance. Mobile phones are very complicated things, but the amount of information required to explain them has a limit. If I knew everything about your mobile phone, I would be able to build one for myself. If I could find a willing supply of cheap, nimble-fingered labour, I could open a factory. The same is true of moon rockets, cars, steam engines, induction cookers, power stations and my six-stage recipe for hummus.

The weather, on the other hand, is complex. There are limits to the predictions we can make about it. There are too many variables, too many interactions and too many feedback loops. Reasonably accurate predictions can be made a few days in advance, but beyond that, forecasts are largely guesses based on what has happened before. Extreme weather events can cause utter devastation, and accurately predicting them a couple of weeks ahead would save countless lives. But alas, such predictions are impossible. The weather occurs within a complex system, and such systems defy prediction. They are also incredibly hard to change. We have flown to the moon and split uranium atoms in two, but we have yet to find a way to stop it from raining.

Obesity is often thought to occur in a complex system, and in many ways it is like the rain. Everyone knows when it is raining. But even the most sophisticated equipment and clever analysis will not be able to tell you exactly why. Several things come together to make it happen, including temperature, pressure, wind speed and humidity. No single cause can be identified, but that doesn't stop people getting wet. So instead of looking for an individual cause, systems science considers the system as a whole, to see if we can make predictions as to how it might behave in the future.

Systems science is a good way of approaching things like meteorology, traffic management, climate change, disease spread and herd immunity. Success in these areas has led many to suggest that viewing obesity in the same way might be a valuable approach. This probably seems a thoroughly sensible idea, but unfortunately, it comes at a cost. Because viewing obesity as a complex system requires us to let go of some fundamental beliefs.

The traditional view has been to look at obesity as a complicated problem, largely caused by people eating too much in response to a limited number of environmental cues. But increasingly, people are starting to frame it as something complex, and therefore harder to control.

Obesity fits many of the criteria that define a complex system:

- Non-linear – Especially in the way the body reacts to decreases in calorie intake, resisting weight loss by dropping resting metabolism. One input does not result in a reliable output.
- Multilayered – It involves systems within systems. For instance, an individual supermarket is one system, but that exists inside a wider retail environment. A house full of

people is a system full of complex interactions, but every house also exists within a wider community.

- Feedback – Many feedback loops exist: food supply shapes people's demand, with manufacturers constantly creating new products and finding clever ways to market them. But equally, demand shapes supply, with supermarkets spending millions monitoring consumer spending patterns, needs and behaviours. If demand for broccoli were to increase, we would quickly see more broccoli appearing in our stores, taking up more shelf-space and becoming easier to spot.
- Chaos – Whenever human behaviour is strongly involved, this is almost a given. A system full of people, especially hungry ones, is unlikely to behave in a predictable manner.

A complex system cannot be understood by measuring all the individual components and adding them together. Equally, changing a single element within a complex system is unlikely to alter the way it functions, unless the system's structure is fundamentally altered. And so, taking this approach, we would not expect the cause of obesity to lie with individual factors, but within the structure of the system as a whole. You can create clever interventions until the cows come home, but if what you are doing does not fundamentally change the system, no one is going to get any thinner. This might explain why, despite hundreds of billions of pounds being spent on obesity research and public health interventions over the past thirty years, people are not losing weight.

Professor Diane Finegood leads a group at Simon Fraser University in Canada, focused on creating models of chronic disease systems. She is widely considered to be one of the world's leading researchers in the field of systems science and obesity. As we start talking, one of the first things she does is to tell me off for carelessly referring to obesity as a complicated problem. It is, she insists, complex. And there's a difference.

Systems thinking and systems approaches recognise that people are not all the same. There are hundreds of genes and other factors affecting them. The food industry is not one thing. It is a collection of heterogeneous organisations. There is lots of interdependence. This means that we need to think about it differently. We can't think about a silver bullet. One thing will not solve the problem.

Unlike the environmental account of obesity, in systems science the role of individuals is not discounted. Despite the huge complexity involved, understanding why people act and appreciating that they have agency is vital. Professor Finegood explains:

> The final arbiter of intake and activity is individuals. You can't take the individual out of the equation, but equally, blaming them won't work. You can't blame people directly for not solving the problem. They respond to their environment, and all you can do is try and help them make better decisions. Although many policy makers now accept that obesity is a complex problem, a lot of strategies stem from the paradigm of individual choice, and are limited in scope.

Professor Terry Huang is director of the Center for Systems and Community Design at the City University of New York. He previously led a programme to integrate systems science approaches into the work of the US National Institutes of Health. On the idea of individuals being responsible for obesity, he told me:

> I don't think individuals are at fault, but I don't want to rob them of their agency. Individual decisions are not made in a vacuum. They are influenced by environmental cues. It is true that most of the time we don't weigh all the pros and cons, and are not fully rational. But in the past we have been guilty of treating people as passive entities. We give them education or impose policy on them. For instance, with the Jamie Oliver school meals, the schools followed the advice, but the parents just brought food to their children. We need to look at what motivates individuals and align what works. For example, we never test interventions that mobilise people politically.

By its very nature it's hard to gather evidence for systems science approaches. But an acceptance that a different approach is required has begun reaching a wider audience.[2] In 2007, the UK government produced the Foresight report, exploring what might be done to tackle obesity.[3] At its heart was a systems map, outlining the many complex structures that can cause someone to gain weight. The map identified a vast network of influence affecting someone's energy balance. It clustered the myriad of factors into six areas: physiology, physical activity, food consumption, food production, individual psychology and social psychology. The result was a huge and terrifyingly complex map, with many connections and feedback loops.

A brief look at the Foresight Map shatters any idea that obesity is as simple as saying 'eat less sugar'. Although not perfect or complete, it is perhaps the clearest illustration of why the simple interventions described at the beginning of this chapter were never likely to have the desired effects. It must seem overwhelming for anyone working in the field of public health or preventative medicine. It could easily lead to the conclusion that because the system is so complex, we might as well give up. After all, you can't stop it raining. But if used correctly, such maps can provide a useful guide to help achieve sustainable change.

Perhaps one of the most frequent criticisms of the systems model of obesity is that it insists upon the importance of engaging with food manufacturers and retailers, clearly a crucial part of the system. This is seen by many as sleeping with the enemy, as a nefarious food industry is often cast as the sole cause of the obesity epidemic. But in systems science, a failure to engage with this vital aspect would be absurd. As Professor Finegood explains:

> You need an appropriate distance, but collaboration with industry is vital for a systems science approach. It is really important to talk to the food industry and work with them. In many cases the industry needs regulation, and often they embrace this, because it maintains a level playing field. There are conspiracy theories, largely because there is a history of public health being influenced by tobacco companies. But it is a different problem from tobacco, and needs a completely different approach. Tobacco was an order of magnitude easier, because you don't need it to live. Even so, it still took a long time to reduce tobacco consumption. Reducing smoking was a case of shifting a deeply held belief that it was cool to be a smoker and easy to obtain cigarettes and to smoke. To change that, we needed to keep on adding interventions to the system. Looking at the tobacco timeline, you might think that taxes were a waste of time. But

just because they didn't create an immediate change does not mean they didn't have an effect. It is often about continuously changing things until you reach a tipping point.

There are many other ways that innovative systems science approaches could help to tackle obesity. As Professor Huang explains: 'It is not about top down interventions, nor bottom up. It is about coordinating between the two. We need to ask different questions to the standard biomedical approach. There is not just one lever that you pull. You need to implement lots of different policies. It becomes both an art and a science.'

Professor Finegood also emphasises the need for interventions to simplify the problem. 'For people to make good decisions,' she says, 'they need to have more capacity than there is complexity. The key is to ask, does this intervention increase complexity? Some food labelling can definitely increase complexity. It makes it harder to make good choices and harder to decide what is healthier. A good system level intervention should always make things simpler.'

Because of the lack of immediate response inherent in many of its interventions, systems science has a problem. As Professor Huang suggests, this turns the creation of public health interventions into an art as well as a science. Policy makers might feel deeply uncomfortable when large amounts of money are at stake. Professor Huang explains: 'The problem is that in society we like to play the blame game. This prevents honest dialogue as to what it will take to create change. What will be the comprehensive strategy to really shift the needle? This lack of coordination has persisted, and we need a new paradigm of a truly multisectoral approach.' Despite a wide acceptance that obesity is a complex problem, there is very little appreciation as to what this actually means in terms of action. But when good practice is adopted, the results can be truly astounding.

AMSTERDAM

In 2013, the deputy mayor of Amsterdam, Alderman Eric van der Berg, decided that firm action was needed to tackle childhood obesity. Like everywhere else in the developed world, rates in the city were rising, especially among the poorest children. Despite spending millions on well-meaning initiatives and programmes to bring rates down, nothing seemed to be working.

With an insight and vision that many modern politicians could learn from, Van der Berg realised that Amsterdam did not require an eye catching advertising campaign, a flash new mobile phone app or an expensively created anti-obesity brand. It needed some leadership. He decided that the many disparate bodies involved in the health and well-being of children should be brought together, and made to work towards a common goal. The focus should not to be on anything as trivial as what children were eating, or even how much exercise they were doing, as these were not the underlying causes. To really tackle childhood obesity, a whole system approach was required. Any successful programme would have to fundamentally change the society children were growing up in, and create a massive cultural shift.

So, instead of publicly shaming fat people, or distributing booklets of recipe ideas, the project focused on the mental, physical and emotional needs of the city's children. Importantly, it concentrated on the most vulnerable children first, helping to narrow rather than widen inequalities. And as this ambitious undertaking required fundamental shifts in attitudes and behaviours, the leaders ensured that there was extensive cross-party, cross-department and cross-sector collaboration.

Perhaps crucially, Van der Berg took the decision that in the initial stages, no new funding would be offered, and no new initiatives launched. Instead, existing projects and resources were audited, and linked together where appropriate. The aim was to create a unified approach to helping the most vulnerable children lead better lives. When gaps were identified, they would be filled by the injection of new funding streams and working groups further down the line.

At no point was a single sector or body held responsible for improvements, as it was recognised that no one in isolation could make a significant difference. And perhaps most controversially, the food industry, including manufacturers and retailers, were extensively involved in the project, and encouraged to find ways of improving food provision throughout the city.

At a community level, the toughest, hardest to reach areas were targeted with specific campaigns, taking into account their environment and needs. There were activity and healthy eating programmes in schools, and bans on local food advertising and sponsorship. Instead of getting rid of school vending machines selling sugary drinks, the decision was taken to allow them, but to ensure there was always access to free tap water as well. Individual children at risk were then educated in the best choices for their needs, and entrusted to make better decisions.

At a city level, there were projects to change the infrastructure, working with urban designers to find ways of encouraging physical activity. Fliers and leaflets were produced to help people adopt healthy behaviours. These focused on eating and drinking well, getting a good night's sleep, and staying active. Notable by its absence was any mention of the word 'obesity'. There were no headless fat bodies. No shaming. No guilt. There were only positive messages, bound by the tag line 'this is how we stay healthy'.

The Amsterdam campaign was in stark contrast to New York a few years previously, when Mayor Michael Bloomberg ran his infamous (failed) attempt to halt the sale of supersize soft drinks. That was accompanied by posters all around the city showing dehumanised images of obese people, many of whom had lost limbs through diabetes. The no-holds-barred approach was clearly designed to evoke feelings of disgust, fear and guilt. But it utterly failed to engage or motivate the public, and the ban was eventually overturned.[4]

However, in Amsterdam, rates of childhood obesity fell by an astonishing 12 per cent in just four years. Perhaps even more remarkably, rates among the poorest children fell by 18 per cent. At the time of writing, Amsterdam is the only place in the developed world where this shameful inequality is narrowing. And the planned twenty-year project has only just begun.[5]

Amsterdam is an example of how a whole system approach can create change, and should be a lesson to anyone interested in combating obesity. But the lesson should not be 'What interventions were effective?' Instead, we should learn that when there is a clear vision, strong leadership and an integrated, system level approach, the seemingly impossible can be achieved. The rain can be stopped.

The behaviours and techniques of the Amsterdam project will doubtlessly be analysed to death, and many attempts will be made to copy and paste them into other cities. But the interventions should be recognised as being unique to Amsterdam. Or perhaps unique to particular communities and neighbourhoods within that city. It would not surprise me at all if some of the most successful changes were specific to the needs of a single child.

There are a few other examples of how system level approaches can prove effective in combatting obesity, or other complex problems. But despite this, and a widespread acceptance that the traditional reductionist approach is not working, there is much resistance to change. We love our old model of disease prevention, the one that has kept us safe and well for so long. We love it for good reason, as it will continue to provide us with answers to complicated problems long into the future. But when it comes to things that are complex, a new approach is sorely needed.

STILL SEARCHING FOR THE BROAD STREET PUMP

Science has proved an extraordinary force when it comes to tackling complicated problems. When John Snow identified the Broad Street pump as the origin of the cholera outbreak, the handle could be removed, and the devastating impact cut short overnight. While it was a complicated problem, it was not a complex one. Making the supply of water in London safe was a huge undertaking, but once we had worked out what needed to be done, it was only a matter of time and resources. It was difficult, but achievable. And now, as a result, most of the world drinks cholera-free water every single day.

Reaching the moon was an immensely complicated task, requiring the greatest minds in the world to collaborate over many years. Tackling diseases such as Ebola is frighteningly complicated, but time and time again science finds a way to outsmart even the most fiendish microorganisms, keeping us a short step ahead in the arms race. Humanity's hive mind and astounding technological innovations keep us safe from harm.

If we can fully understand and categorise the task, observe all the variables and find exactly where the pump handle is located, we can solve the most complicated problems life throws at us. Even AIDS, one of the most challenging health problems of all, can be faced down. With the will and resources, AIDS can be defeated. And with the ingenuity and limitless compassion of humanity, I genuinely believe that one day it will be.

But as the saying goes, you've got to die of something. When all the complicated problems have been solved, what remain are the

complex ones. Obesity. Type 2 diabetes. Cancer. Mental health problems. Suicide. Global warming. Extreme weather events. Terrorism.
Sexual abuse. Inequality. All of these are expressed in complex systems.
All resist the effects of simple top-down intervention. All will change
and mutate as you attempt to prevent them. Trying to combat a complex
problem is like fighting a multiheaded hydra. Cut off one head and
another will grow, sometimes stronger, meaner and harder to fight than
the last. Whenever you act upon a complex system, you never really
know what the reaction will be. You can imagine. You can make good
guesses. You can probably design a flashy algorithm to make those
guesses more informed. But you never really know.

Understanding obesity is like trying to understand the economy, but
with a layer of bioscience and social psychology on top, and a huge
amount of prejudice, stigma, vested interests and political pressures
thrown in for good measure. Good luck trying to develop one intervention to change it. And good luck trying to test that intervention in the real
world, because I can near guarantee that it will not work in the way that
you think it will.

Try taxing sugar-sweetened drinks, a sensible idea touted by many
as a possible solution. Maybe as the price rises, people will buy fewer
of them. But that change will not occur in isolation. Other behaviours
will be affected. Some of the calories will be compensated for. The
system will move in ways that cannot be predicted, and not all of them
will be positive. Things will vary greatly across different population
groups. People's metabolisms will resist any small changes in calorie
intake, and stop them from losing weight.

Perhaps everyone will start buying cheap alcohol instead. Maybe
they will start buying illegal or grey market drinks that have even more
sugar than the ones being replaced. Or they might just buy the drinks at
the higher price, making them unable to spend the money elsewhere. So
they will buy fewer vegetables. Or school books. Or heating.

But maybe the tax will also change the attitude of society to these
drinks. Or encourage manufacturers to reformulate or reduce portion
size. Or stimulate the development of new taste technologies to make
better-tasting low-calorie drinks. But then again, maybe they will swap
in a new ingredient that has a completely different impact on health.
Maybe it will alienate the soft drink companies, meaning that they will
concentrate their marketing resources on vulnerable individuals in
developing countries, where regulation is not as tight.

Even when we deliver the most tailored and sophisticated programmes possible, sometimes nothing happens. Occasionally, we might make things worse in some unexpected way. The conventional model of medical research will sometimes deem the best interventions useless, and abandon them in a fruitless hunt for a Broad Street pump. And so we are left desperate, hunting for certainty in a complex, uncertain world.

If it is not exercise, it must be carbs. Or processed food. Or trans fat, saturated fat, drugs, chemicals, sleep, diet culture, portion size, disease, the environment, poverty, inequality, loneliness, isolation, food marketing, fructose, neoliberalism, trauma, binge eating disorder, microbiome disruption, stress, built environments, cars, stigma, screen time, laziness, addiction, assortative mating, maternal age, ambient temperature or eating.

If we look hard through a conventional lens, we can probably prove that obesity is caused by none of these things. But in all likelihood it is actually affected by all of them, and a thousand other factors I haven't been able to cover in this book. All acting together, pushing and pulling us in different directions like a three-dimensional tug of war. Remove any one of these factors, and obesity might still occur. But that does not mean that they did not contribute. That is just how complex systems work.

And within the system sits us, with our own astounding complexity and individuality. Science barely understands the physiological processes that control our appetite and define the set point weight that our bodies fight to maintain. On top of that, we have our behaviour, controlled by our brains, the most complex object in the known universe. We have the capacity to exert conscious control over our appetite, yet whether we decide to or not is an extraordinarily complex decision, impossible to predict in any meaningful way. It will be influenced by our history, our values, our beliefs and our desires. It is as individual as our fingerprint.

It is worth remembering that obesity might just be the inevitable result of free access to food. Maybe it is not a sign of moral failing, just an artefact of the human condition. Who are we to say what weight is normal, or what rate of obesity is natural? Maybe right now is the only time in the history of humanity that we are seeing ourselves as we really are. And why should our cultural and aesthetic expectations align with our biology? People used to think white skin was sexy because it meant you didn't have to work in a field, then people liked tanned skin because it meant you could afford holidays. Our aesthetic and sexual preferences are not timeless, they are mutable and conditioned, based on our shared culture.

If you really embrace complexity, then the most unsettling thing of all is having to accept that we probably already know the answers. Sometimes our gold standard of randomised controlled trials might not show success, even for the most beautifully designed interventions. We need to change things from the top down, but also from the bottom up. And from the left and right as well.

THE SOLUTION

I promised the solution to the problem of obesity at the beginning of this chapter. Well, here it is.

Obesity is a system level problem, so it requires a system level solution. When individuals are obese and suffering from health difficulties, they need professionals to work closely with them and help them adopt better behaviours in a realistic and sustainable way. The focus should be on improving their health, even if that does not mean reducing their weight. Interventions need to occur over the long term, with constant, dedicated support and encouragement. If things are not working well, the professionals involved in treatment should alter the approach accordingly. People should be given realistic plans, tailored to their needs and requirements. Any changes to people's diets should fit into the context of their lives, accounting for their preferences, resources and lifestyle.

Dietitians should work in collaboration with medical doctors, psychologists and fitness professionals to find the fundamental problems underlying someone's weight gain. For some, weight might be so hard to shift that the best approach would be to embrace more positive health behaviours. For others, improved dietary quality and psychological support might make a huge difference, perhaps making them less fat. In a few rare cases, surgery or pharmaceutical treatments might be the best option. For others, self-acceptance and access to exercise facilities might do the trick. Above all, this help should be targeted first at those who need it the most. People leading difficult lives. Those on low incomes, in deprived areas. We should always aim to be reducing inequalities, not perpetuating and widening them.

At the community level, public health professionals should work closely with local residents, conducting surveys and focus groups to assess the problems people are facing. Some areas might need better exercise facilities. We might need to address poor-quality housing, a lack of work opportunities, low educational attainment, long-term unemployment and low pay. There might be problems of social stigma being perpetuated by some residents, which could be addressed with education campaigns. Maybe some areas would benefit from cookery classes, or the opening of takeaways selling healthier options. This could be encouraged by lowering business rates and local taxation laws for these establishments, or perhaps engaging nutritionists to work closely with existing businesses, creating financial incentives to sell healthier foods. Maybe street food markets could be encouraged to visit the area, creating hubs where residents might encounter more diverse and interesting cuisines. Everything would depend on the specific needs of individual communities, and local public health workers would be empowered to make decisions based on the best options for the people they serve. The initial focus would be on the poorest communities, widening the support out only when lasting changes have been made in these areas.

At the national level, the greatest thing that any government could do would be to tackle financial inequality. To give voice and power to the millions living in poverty. To instil their lives with meaning and let them feel valued once again. To rebuild shattered post-industrial communities and bring in businesses that will find use for the huge untapped labour resource. And when that is done, perhaps they might combat the vicious stigma that fat people face every day. To educate the public that this is no longer acceptable. To shout loudly that this abuse is as damaging, illogical and stupid as any other form of prejudice.

The corporations that manufacture and sell foods need to be involved, working with local communities to improve the food on offer, and encouraging the purchase of a variety of interesting, nutritious products. Companies could be ranked and graded by their involvement in this effort, and levers put in place to ensure that this ranking strongly affects share price and profitability. The worst performing companies would be named, shamed and fined, the best praised and supported with new investment.

Investment in innovations that encourage people to eat a healthy variety of foods would create big tax advantages. Promotional mechanics in retail stores that push people towards better choices would be heavily incentivised. Supermarket consumer data would be opened up to academics, to help them develop strategies to encourage better shopping habits.

All that stuff, plus a load more. Everything would need to be done together, without stopping every step of the way to perform a randomised controlled trial. Even if we did manage to do all that, I have no idea if we would be any thinner. I don't even know if we would stop gaining weight. But I can guarantee we would be happier and healthier, and that the world would be a better place to live. Which presumably is why we are doing any of it in the first place.

The solution to obesity is right in front of us. We could start solving it today. But instead we continue our endless search for a cure using a tried and tested process that solved so many complicated problems in the past. So we argue about carbs, fats, diets and food environments. We blame conspiracies, corporations and moral decline. We search for a simple cause in sleep, stress or the bacteria in our gut.

If we really want to solve the problems of obesity, we need to accept that there is no single cure, because there was never a single cause. The real solution lies in years of hard work at individual and community levels. And perhaps for us all to accept some difficult truths about the type of world that we live in now.

For it to really work, we would all need to play a part. We would need to start accepting people as they are. We would need to call out prejudice. Develop a better relationship with food and our bodies. Lose some of our guilt around eating. Embrace more variety into our diets. Remember that the truly important thing about food is the way that it brings us together.

We should take some time for lunch, and invite people along whenever we can. Share good food. Try new stuff. Reconnect with eating, and find foods that bring us joy. Include others every time we cook. Rebuild our communities. Find some shared purpose. Take more time to run, jump and play. And try to get some sleep.

EPILOGUE:
WHAT TO DO IF YOU ARE FAT

I hope that I have given you something to think about, and perhaps some useful information to help you in your life. The intention of this book was to help people understand more about obesity, and accept that it is a deeply complex problem. Although accepting complexity is the hard road, pretending it is simple only defers a greater disappointment.

Of course, there are still those who tell us these problems are simple and easily solved. Complex problems that confound conventional scientific thinking have become the new targets of charlatans. Cancer, autism, stress, fatigue and mental health are all complex problems without simple solutions, which is exactly why the wolves of pseudoscience tear at the flesh of sufferers, drawing them in with false declarations of certainty. Homeopathy rarely offers solutions to smallpox or cholera, because medical science nailed those long ago. But where complexity confounds medicine, leaving easy answers in short supply, many are drawn to false prophets.

So it is with obesity. Countless charlatans sell you their diets, or offer up conspiracy theories to explain obesity's rise. When it comes to the diet industry, purveyors of false hope are thoroughly legitimised, sometimes even prescribed by doctors. They draw us in with certainty, and demand we buy their products, promising hidden secrets will be revealed. If there is any single message to take from this book, it is that these people don't have anything close to a monopoly on the truth.

But this does beg the question: what exactly should people do if they are fat, and don't want to be?

Sadly, however much we want a solution that is both simple and effective, it does not exist. Just as a special diet will never make me taller, or alter my weirdly shaped right ear, there is unlikely to be a dietary change that will sustainably make you thin. It may be possible for you to live at a slightly lower weight in a healthy way, but beyond a window of a few kilograms, anything else is unlikely.

Clearly, I am just a chef, and if you really want advice about your diet and health you should speak to a doctor or dietitian (and preferably one who knows you, not some random social media type). I hope that by now it is obvious that I fully admit my privilege, and try hard to appreciate the difficulties fat people face in the world. But if you really want to lose weight, I think you need to ask yourself why.

It might be that you want to be thinner because you think that will make you more attractive. Many anti-diet advocates are keen to dismiss this as needless vanity, but most people want to look as nice as they can, and should not feel ashamed that they do. But any desire for short-term weight loss should be weighed against the known risks, and lack of long-term success. And any belief that losing weight is the only thing that you can do to make yourself more attractive is well worth challenging.

If you are concerned that people will not find you attractive as you are right now, then it might be worth thinking about what, beyond your weight, might change that. Anyone choosing a partner based on their BMI is not a great catch. So maybe think about the sort of person that you want to be, and the many aspects you can control far more easily and sustainably than your weight. There are a million things that define how attractive someone is. It is a fucked up society that believes weight is the only one that matters.

If you want to lose weight because you are worried about your health, that is certainly a valid concern. Type 2 diabetes, cardiovascular disease and cancer are all devastating conditions, and we should do everything we can to reduce our risk. Thankfully, even if you do not lose weight, there are several things that you can do, all with good evidence to support their effectiveness. If we set aside the drastic intervention of bariatric surgery, which is both effective and risky, you could stop smoking and eat a more varied diet with plenty of fruits, vegetables and fibre. You could avoid trans fats, though they are already being phased out in the UK and the US. You could try to get some exercise. You could attempt to sleep better. You could take action to reduce the stress levels in your life, and think about addressing any mental health concerns you might be worried about. If mobility is a problem, consider what steps might help you become more mobile.

Are there strengthening exercises or changes to your life that might make things easier? And if health really is your main concern, one thing you should definitely avoid is a strict weight loss diet, and the endless cycling and regain that almost always results.

It might be that if you do all those things, especially concentrating on your sleep, mental well-being and quality of your diet, you will lose a bit of weight. If you exercise over the long term, it might help to adjust your set point weight. Equally, it might not, but you might find instead that you are happier and healthier.

And for anyone not interested in weight loss, I would urge you to do exactly the same things, because they are important for us all. But I would also ask just two more favours, because I think they might make the world a better place.

The first is to stop judging people by how they look. We all know that weight has absolutely nothing to do with someone's moral character or worth, so we need to stop behaving as if it does.

Finally, try to eat together a little more. Share some sandwiches in the park with an old friend. Get a takeaway to eat with your family. Take five minutes for coffee and biscuits with a neighbour. Buy ice creams for your children, and enjoy one yourself. Cook lunch for your workplace, and insist everyone takes a few minutes to sit down and enjoy it together.

The 'obesity epidemic', and the moral panic surrounding it, has caused a great deal of damage, affecting people in all manner of ways. But perhaps worst of all, it has deeply damaged our relationship with food, something that desperately needs mending if we are to move forward. In all the guilt, dieting, blame and restriction, we seem to have forgotten the most important things that food can do for us. For although it is neither magic nor medicine, when food helps us form new bonds and brings us closer to those we love, it can truly transform our lives.

ACKNOWLEDGEMENTS

A great many people helped in the writing and research for this book, giving up their valuable time. I am forever grateful to all of them. Many are quoted in the text, but others contributed too, helping to expand and develop my knowledge of often complex subjects. In addition to everyone quoted, I would like to thank Stephan Guyenet, Jane Ogden, Chris Snowden, Harry Rutter, Sharon Noonan-Gunning, Nadia Mehdi, Angela Meadows, Sally Marlow, Lucy Aphramor, Mike Gibney and Phillipe Vandenbroeck. Apologies if I have missed anyone out.

Special thanks to Laura Thomas, Helen West and Judy Swift, all of whom put up with many stupid questions, and showed great patience and understanding. You helped shape my arguments, and I will be interested to find out if you agree with them.

Despite all this help, late in 2017 the task of writing this book began to overwhelm me. Like any good superhero, Captain Science, my anonymous collaborator, came to the rescue. Without her, I genuinely do not think I would have reached the end. I thank her for her patience, understanding and razor-sharp mind. With her in the world, pseudoscientists and charlatans stand little chance. Be afraid.

Thanks to everyone at Oneworld, particularly Alex Christofi and Kate Bland, who have shown huge belief and support over the past couple of years. And thanks to Alexandra Cliff and everyone at PFD. None of this would be possible without you.

I am blessed with a family who understands and supports me in everything I do, however foolish. This book required me to spend many hours in a shed at the bottom of the garden, locked away from my responsibilities as a husband and father. I can only apologise, and promise that I will make up for all the times that I was not there for you both. Although the first book was hard, this one cost a bit too much. That will not happen again.

Ellie inspires me every day, with bravery and determination that I can only aspire to. I am already immensely proud of everything she has achieved, and her story is only just beginning.

And Mrs Angry Chef will forever be my hero, mostly just for putting up with me. Everything I achieve is made possible by her. And every success belongs to us both.

NOTES

INTRODUCTION

1. OECD, *Health at a Glance 2017: OECD Indicators*

PART I: THE MODERN EPIDEMIC
1 WHY DO WE GET FAT? PART I

1. Speakman, et al., 'Set points, settling points and some alternative models: theoretical options to understand how genes and environments combine to regulate body adiposity', *Disease Models & Mechanisms*, 4 (6), pp. 733–45, http://doi.org/10.1242/dmm.008698; 'Changes in Waist Circumference among German Adults over Time – Compiling Results of Seven Prospective Cohort Studies', *Obesity Facts: The European Journal of Obesity*, 9(5), pp. 332–43, https://doi.org/10.1159/000446964; M. Ebrahimi-Mameghani et al., 'Changes in Weight and Waist Circumference over 9 years in a Scottish population', *European Journal of Clinical Nutrition*, 2008, 62, pp. 1208–14, https://www.nature.com/articles/1602839
2. S. S. Gidding, R. L. Leibel, S. Daniels, M. Rosenbaum, L. Van Horn and G. R. Marx, 'Understanding Obesity in Youth – A Statement for Healthcare Professionals From the Committee on Atherosclerosis and Hypertension in the Young of the Council on Cardiovascular Disease in the Young and the Nutrition Committee, American Heart Association', *Circulation*, http://circ.ahajournals.org/content/94/12/3383
3. A. Kroke, F. Manz, M. Kersting, T. Remer, W. Sichert-Hellert, U. Alexy and M.J. Lentze, 'The DONALD Study. History, current status and future perspectives.' *European Journal of Nutrition*, 2004, February;43 (1), pp. 45–54, https://www.ncbi.nlm.nih.gov/pubmed/14991269
4. T.V.E Kral, R.I. Berkowitz, A.J. Stunkard, V.A. Stallings, D.D. Brown and M.S. Faith, 'Dietary energy density increases during early childhood irrespective of familial predisposition to obesity: results from a prospective cohort study', *International Journal of Obesity*, 31, pp. 1061–7, https://www.nature.com/articles/0803551
5. D.M. Garner and S.C. Wooley, *Clinical Psychology Review*, Volume 11, Issue 6,1991, pp. 729–80, https://www.sciencedirect.com/science/article/pii/027273589190128H#
6. A.M. Prentice and S.A. Jebb, 'Obesity in Britain: gluttony or sloth?' *British Medical*

Journal, August 1995, 12:311 (7002), pp. 437–9, https://www.ncbi.nlm.nih.gov/pubmed/7640595

7. J. Raisborough, *Fat Bodies, Health and the Media* (London: Palgrave Macmillan, 2016)
8. G.R. Hervey, 'A hypothetical mechanism for the regulation of food intake in relation to energy balance', *Proc. Nutr. Soc.*, 1969, September, 28(2), pp. 54A–55A, https://www.ncbi.nlm.nih.gov/pubmed/5353342
9. D. Benton and H. A. Young, 'Reducing Calorie Intake May Not Help You Lose Body Weight', *Perspectives on Psychological Science*, 2017, September, 12(5), pp. 703–14, https://www.ncbi.nlm.nih.gov/pmc/articles/PMC5639963/
10. M. G. Myers, Jr.,R. L. Leibel, R. J. Seeley and M. W. Schwartz, 'Obesity and Leptin Resistance: Distinguishing Cause from Effect', *Trends in Endocrinology Metabolism*, 2010, 21(11), pp. 643–51, https://www.ncbi.nlm.nih.gov/pmc/articles/PMC2967652/

2 WHY DO WE GET FAT? PART II

1. M., Kojima, H. Hosoda, Y. Date, M. Nakazato, H. Matsuo, K. Kangawa, 'Ghrelin is a growth-hormone-releasing acylated peptide from stomach', *Nature*, 9 December 1999, 402,656–60,https://www.nature.com/articles/45230
2. 'Set points, settling points and some alternative models: theoretical options to understand how genes and environments combine to regulate body adiposity', *Disease Models and Mechanisms*, November 2011, 4(6), pp. 733–45, https://www.ncbi.nlm.nih.gov/pmc/articles/PMC3209643/
3. S.H. Lockie and Z.B. Andrews, 'The hormonal signature of energy deficit: Increasing the value of food reward', *Molecular Metabolism*, November 2013, 2(4), pp. 329–36, https://www.ncbi.nlm.nih.gov/pmc/articles/PMC3854986/
4. B. Perry and Y. Wang, 'Appetite regulation and weight control: the role of gut hormones', *Nutrition and Diabetes*, 2012, 2,e26, https://www.nature.com/articles/nutd201121
5. Jansson et al., 'Body weight homeostat that regulates fat mass independently of leptin in rats and mice', *PNAS*, http://www.pnas.org/content/115/2/427
6. H.R. Berthoud, H. Münzberg, C.D. Morrison, 'Blaming the Brain for Obesity: Integration of Hedonic and Homeostatic Mechanisms', *Gastroenterology*, May 2017, Volume 152, Issue 7, pp. 1728–38, https://www.gastrojournal.org/article/S0016-5085(17)30143-9/fulltext
7. H.R. Berthoud, 'The vagus nerve, food intake and obesity', *Regulatory Peptides*, 7 August 2008, Volume 149, Issues 1–3, pp. 15-25, https://www.sciencedirect.com/science/article/pii/S0167011508000621?via%3Dihub
8. Chin Jou, '*The New England Journal of Medicine*2014, 370, pp. 1874–77, https://www.nejm.org/doi/full/10.1056/NEJMp1400613
9. C. von Loeffelholz, 'https://www.ncbi.nlm.nih.gov/books/NBK279077/
10. McArdle et al., *Exercise Physiology: Nutrition Energy and Human Performance*, eighth international edition (Baltimore: Lippincott Williams and Wilkins, 2014)

3 WHY DON'T WE JUST EAT LESS?

1. Sumithran et al., '*The New England Journal of Medicine* 27 365, pp. 1597–04, https://www.nejm.org/doi/full/10.1056/NEJMoa1105816

2. Rosenbaum et al., 'Energy Intake in Weight-Reduced Humans', *Brain Research*, 2 September 2010, Volume 1350,95–102, https://www.sciencedirect.com/science/article/pii/S0006899310012667?via%3Dihub

3. D.A. Levitsky and L. DeRosimo, 'One day of food restriction does not result in an increase in subsequent daily food intake in humans', *Physiology and Behavior*, 30 March 2010, Volume 99, Issue 4,pp. 495–9, https://www.sciencedirect.com/science/article/abs/pii/S0031938409004077

4. J.H. Lavin, S.J. French and N.W. Read, 'The effect of sucrose- and aspartame-sweetened drinks on energy intake, hunger and food choice of female, moderately restrained eaters', *International Journal of Obesity Related Metabolic Disorders*, January 1997, 21(1), pp. 37–42, https://www.ncbi.nlm.nih.gov/pubmed/9023599; Benton and Young, 'Reducing Calorie Intake May Not Help You Lose Body Weight', pp. 703–14

5. M. Garnder and S.C. Wooley, 'Confronting the failure of behavioral and dietary treatments for obesity', *Clinical Psychology Review*, 1996, 11 (6), pp. 729–80, https://www.sciencedirect.com/science/article/pii/027273589190128H#

6. Mann et al., 'Medicare's search for effective obesity treatments: diets are not the answer', *American Psychologist*, 2007, (3), pp. 220–33, http://janetto.bol.ucla.edu/index_files/Mannetal2007AP.pdf

7. Ibid; K. Strohacker, K.C. Carpenter and B.K. McFarlin, 'Consequences of Weight Cycling: An Increase in Disease Risk?', *International Journal of Exercise Science*, 2009, 2(3), pp. 191–201, https://www.ncbi.nlm.nih.gov/pmc/articles/PMC4241770/

8. Johns et al., 'Weight change among people randomized to minimal intervention control groups in weight loss trials', *Obesity* (Silver Spring), April 2016, 24(4), pp. 772–80, doi: 10.1002/oby.21255, https://www.ncbi.nlm.nih.gov/pubmed/27028279; Hartman-Boyce et al., 'Behavioural weight management programmes for adults assessed by trials conducted in everyday contexts: systematic review and meta-analysis', *Obesity Review*, November 2014, 15(11), pp. 920–32, https://www.ncbi.nlm.nih.gov/pubmed/25112559

9. Parretti et al., 'Clinical effectiveness of very-low-energy diets in the management of weight loss: a systematic review and meta-analysis of randomized controlled trials', *Obesity Review*, March 2016, 17(3), pp. 225–34, https://www.ncbi.nlm.nih.gov/pubmed/26775902

10. Field et al., 'Prospective Association of Common Eating Disorders and Adverse Outcomes', *Pediatrics*, August 2012, 130(2), pp. e289–e295, http://pediatrics.aappublications.org/content/early/2012/07/11/peds.2011-3663

11. J. Lebow, L.A. Sim and L.N. Kransdorf, 'Prevalence of a history of overweight and obesity in adolescents with restrictive eating disorders', *Journal of Adolescent Health*, January 2015, 56(1), pp. 19–24, https://www.ncbi.nlm.nih.gov/pubmed/25049202

12. Neumark-Sztainer et al., 'Dieting and unhealthy weight control behaviors during adolescence: Associations with 10-year changes in body mass index', *Journal of Adolescent Health*, January 2012, 50(1), pp. 80–86, https://www.ncbi.nlm.nih.gov/pubmed/22188838

13. A.E. Field, P. Aneja, S. Bryn Austin, L.A. Shrier, C. de Moor and P. Gordon-Larsen, 'Race and Gender Differences in the Association of Dieting and Gains in BMI among Young Adults', *Obesity*, February 2007, Volume 15 No. 2, https://onlinelibrary.wiley.com/doi/pdf/10.1038/oby.2007.560

14. W. Wayt Gibbs, 'Obesity: An Overblown Epidemic?', *Scientific American*, 1 December 2006, https://www.scientificamerican.com/article/obesity-an-overblown-epidemic-2006-12/
15. M.R. Lowe, 'Dieting: proxy or cause of future weight gain?', *Obesity Review*, February 2015, 16 Suppl 1, pp.19–24, https://www.ncbi.nlm.nih.gov/pubmed/25614200; A.J. Hill, 'Does dieting make you fat?', *British Journal of Nutrition*, August 2004, 92 Suppl 1:S15–8, https://www.ncbi.nlm.nih.gov/pubmed/15384316
16. A.J. Hill, 'Does dieting make you fat?', *British Journal of Nutrition*, August 2004, 92 Suppl 1:S15–8, https://www.ncbi.nlm.nih.gov/pubmed/15384316
17. S.E. Saarni, A. Rissanen, S. Sarna, M. Koskenvuo and J. Kaprio, 'Weight cycling of athletes and subsequent weight gain in middle age', *International Journal of Obesity* (London), November 2006, 30(11), pp. 1639–44, https://www.ncbi.nlm.nih.gov/pubmed/16568134
18. K.H. Pietiläinen, S.E. Saarni, J. Kaprio A. Rissanen, 'Does dieting make you fat? A twin study', *International Journal of Obesity*, 9 August 2011, 36,456–64, https://www.nature.com/articles/ijo2011160
19. A.G. Dulloo, J. Jacquet, J.L. Miles-Chan, Schutz, 'Passive and active roles of fat-free mass in the control of energy intake and body composition regulation', *European Journal of Clinical Nutrition*, 2017, V71,353–7, https://www.nature.com/articles/ejcn2016256
20. L. Bacon and L. Aphramor, 'Weight Science: Evaluating the Evidence for a Paradigm Shift', *Nutrition Journal* https://nutritionj.biomedcentral.com/articles/10.1186/1475-2891-10-9
21. *Clinical Practice Guidelines for the Management of Overweight and Obesity in adults, Adolescents and Children in Australia*, Australian Government's National Health and Medical Research Council, 2013, https://www.nhmrc.gov.au/_files_nhmrc/publications/attachments/n57_obesity_guidelines_140630.pdf

4 HOW FAT ARE WE?

1. C.L. Ogden, M.D Carroll, L.R. Curtin, M.A. McDowell, C.J. Tabak, K.M. Flegal, 'Prevalence of overweight and obesity in the United States, 1999–2004', *Journal of the American Medical Association*, April 2006, 5, 295(13), pp. 1549–55, https://www.ncbi.nlm.nih.gov/pubmed/16595758
2. B.M. Popkin and Gordon-Larsen, 'The nutrition transition: worldwide obesity dynamics and their determinants', *International Journal of Obesity*, 2004, 28, S2–S9, https://www.nature.com/articles/0802804
3. B. Caballero, 'The Global Epidemic of Obesity: An Overview', *Epidemiologic Reviews*, 1 January 2007, Volume 29, Issue 1, pp. 1–5, https://academic.oup.com/epirev/article/29/1/1/444345
4. M. Gard and J. Wright, *The Obesity Epidemic: Science, Morality, and Ideology*
5. H. Blackburn and D. Jacobs, Jr., 'Commentary: Origins and evolution of body mass index (BMI): continuing saga', *International Journal of Epidemiology*, June 2014, Volume 43, Issue 3, https://academic.oup.com/ije/article/43/3/665/2949550
6. L.J. Lloyd, S.C. Langley-Evans and S. McMullen, 'Childhood obesity and risk of the adult metabolic syndrome: a systematic review', *International Journal of Obesity*, 2012, 36, pp. 1–11, https://www.ncbi.nlm.nih.gov/pmc/articles/PMC3255098/pdf/ijo2011186a.pdf
7. E. Gill, 'This mum is furious after she got a "laughable" letter claiming her

11-year-old son is overweight', *Manchester Evening News*, 3 July 2017, https://
www.manchestereveningnews.co.uk/news/greater-manchester-news/mum-furious-
after-laughable-letter-13275910

8. R. Bishop, ' Mum furious after "fit and healthy" veggie-loving son aged 5 is branded
"overweight" in "fat-shaming letter" from school', *Daily Mirror*, 7 May 2017, https://
www.mirror.co.uk/news/uk-news/mum-furious-after-fit-healthy-10376309

9. C. Griffiths, P. Gately, P.R. Marchant and C.B. Cooke, 'Cross-Sectional
Comparisons of BMI and Waist Circumference in British Children: Mixed
Public Health Messages', *Obesity*, June 2012, Volume 20, Issue 6, pp. 1258–60,
https://onlinelibrary.wiley.com/doi/full/10.1038/oby.2011.294

10. R.C. Davey and R. Stanton, 'The obesity epidemic: too much food for thought?',
British Journal of Sports Medicine, 2004, Volume 38, Issue 3, pp. 360–63, http://
bjsm.bmj.com/content/38/3/360

11. N.A. Schvey, R.M. Puhl and K.D. Brownell, 'The impact of weight stigma on
caloric consumption', *Obesity* (Silver Spring), October 2011, 19(10):, pp.
957–62, https://www.ncbi.nlm.nih.gov/pubmed/21760636

12. K.M. Flegal, M.D. Carroll, R.J. Kuczmarski and C.L. Johnson, 'Overweight and
obesity in the United States: prevalence and trends, 1960–1994', *International
Journal of Obesity Related Metabolic Disorders*, January 1998, 22(1), pp. 39–47,
https://www.ncbi.nlm.nih.gov/pubmed/9481598

13. Ibid. pp. 235–41

14. R. Sturm, 'Increases in clinically severe obesity in the United States, 1986–
2000', *Archives of Internal Medicine*, October 2003, https://www.ncbi.nlm.nih.
gov/pubmed/14557211

15. L. Bartoshuk, 'The "Obesity Epidemic"', *Observer*, Association for Psychological
Science, January 2010, Volume 23, Issue 1, https://www.psychologicalscience.org/
observer/the-obesity-epidemic

16. J. Komlos and M. Brabec, 'The trend of BMI values of US adults by deciles, birth
cohorts 1882–1986 stratified by gender and ethnicity', *Economics and Human
Biology*, 2011, 9 pp. 234–50, http://user37685.vs.easily.co.uk/wp/wp-content/
uploads/2013/10/komlos-brabec-2011.pdf

17. Bartoshuk, 'The "Obesity Epidemic"'

18. K.M. Flegal, R.P. Troiano, E.R. Pamuk, R.J. Kuczmarski and S.M. Campbell,
'The influence of smoking cessation on the prevalence of overweight in the
United States', *New England Journal of Medicine*, November 1995, 2, 333(18),
pp. 1165–70, https://www.ncbi.nlm.nih.gov/pubmed/7565970?dopt=Abstract&
holding=npg

19. C. Courtemanche, R. Tchernis and B. Ukert, 'The Effect of Smoking on Obesity:
Evidence from a Randomized Trial', National Bureau of Economic Research
Working Paper No. 21937, June 2016, http://www.nber.org/papers/w21937

4 HOW UNHEALTHY IS IT TO BE FAT?

1. K.M. Flegal, Graubard, D.F. Williamson and M.H. Gail, 'Excess deaths associated with underweight, overweight, and obesity', *Journal of the American Medical
Association*, April 2005, 20, 293(15), pp. 1861–7, https://www.ncbi.nlm.nih.gov/
pubmed/15840860/

2. D.B. Allison, M.S. Faith, M. Heo, D. Townsend-Butterworth and D.F.
Williamson, 'Meta-analysis of the effect of excluding early deaths on the

estimated relationship between body mass index and mortality', *Obesity Research*, July 1999, 7(4), pp. 342–54, https://www.ncbi.nlm.nih.gov/pubmed/10440590

3. A.H. Mokdad, J.S. Marks, D.F. Stroup and J.L. Gerberding, 'Actual causes of death in the United States, 2000', *Journal of the American Medical Association*, March 2004, 10, 291(10), pp. 1238–45, https://www.ncbi.nlm.nih.gov/pubmed/15010446

4. http://archive.sciencewatch.com/ana/st/obesity2/10sepObes2Fleg1/

5. K.M. Flegal, B.K. Kit, H. Orpana and B.I. Graubard, 'Association of all-cause mortality with overweight and obesity using standard body mass index categories: a systematic review and meta-analysis', *Journal of the American Medical Association*, January 2013, 2, 309(1), pp. 71–82, https://www.ncbi.nlm.nih.gov/pubmed/23280227

6. Global BMI Mortality Collaboration, 'Body-mass index and all-cause mortality: individual-participant-data meta-analysis of 239 prospective studies in four continents', *The Lancet*, August 2016, 20, 388(10046), pp: 776–86, https://www.ncbi.nlm.nih.gov/pubmed/27423262

7. K.M. Flegal and J.P.A. Ioannidis, 'A meta-analysis but not a systematic review: an evaluation of the Global BMI Mortality Collaboration', *Journal of Clinical Epidemiology*, August 2017, 88, pp. 21–9, https://www.ncbi.nlm.nih.gov/pubmed/28435099

8. Duan et al., 'Body mass index and risk of lung cancer: Systematic review and dose-response meta-analysis', *Scientific Reports*, 5, Article number: 16938, https://www.nature.com/articles/srep16938

9. K.F. Petersen and G.I. Shulman, 'Etiology of insulin resistance', *American Journal of Medicine*, May 2006, 119, (5 Suppl 1), pp. S10–16, https://www.ncbi.nlm.nih.gov/pubmed/16563942

10. GBD, 'Disease and Injury Incidence and Prevalence Collaborators – Global, regional, and national incidence, prevalence, and years lived with disability for 310 diseases and injuries, 1990–2015: a systematic analysis for the Global Burden of Disease Study 2015', *The Lancet*, 8 October 2016, 388(10053), pp. 1545–1602, https://www.ncbi.nlm.nih.gov/pmc/articles/PMC5055577/

11. S. Smyth and A. Heron, 'Diabetes and obesity: the twin epidemics', *Nature Medicine*, January 2006, V12,75–80, https://www.nature.com/articles/nm0106-75

12. P. Campos, A. Saguy, P. Ernsberger, E. Oliver and G.A. Gaesser, 'The epidemiology of overweight and obesity: public health crisis or moral panic?' *International Journal of Epidemiology*, 1 February 2006, Volume 35, Issue 1, pp. 55–60, https://academic.oup.com/ije/article/35/1/55/849914

13. J. Kaur, 'A comprehensive review on metabolic syndrome', *Cardiology Research and Practice*, 11 March 2014, https://www.ncbi.nlm.nih.gov/pubmed/24711954

14. 'The epidemiology of overweight and obesity: public health crisis or moral panic?', pp. 55–60

15. F.B. Hu, 'Obesity and mortality: watch your waist, not just your weight', *Archives of Internal Medicine*, 2007, 167(9), pp. 875–6, https://jamanetwork.com/journals/jamainternalmedicine/article-abstract/412348; C. Kragelund and T. Omland, 'A farewell to body-mass index?', *The Lancet*, 5 November 2005, Volume 366, No. 9497, pp. 1589–91, https://www.thelancet.com/journals/lancet/article/PIIS0140-6736(05)67642-8/abstract

16. G.A. Gaesser, S.S. Angadi and B.J. Sawyer, 'Exercise and diet, independent of weight loss, improve cardiometabolic risk profile in overweight and obese

individuals', *The Physician and Sportsmedicine*, May 2011, 39(2), pp. 87–97, https://www.ncbi.nlm.nih.gov/pubmed/21673488

17. A.R. Sutin, Y. Stephan and A. Terracciano, 'Discrimination and Risk of Mortality', *Psychological Science*, 29 September 2015. http://journals.sagepub.com/doi/pdf/10.1177/0956797615601103

18. E. Robinson, A. Haynes, A.R. Sutin and M. Daly, 'Telling people they are over-weight: helpful, harmful or beside the point?', *International Journal of Obesity* (London), August 2017, 41(8), pp. 1160–61, https://www.ncbi.nlm.nih.gov/pubmed/28785104

19. A. Haynes, I. Kersbergen, A. Sutin, M. Daly and E. Robinson, 'A systematic review of the relationship between weight status perceptions and weight loss attempts, strategies, behaviours and outcomes', *Obesity Reviews*, March 2018, 19(3), pp. 347–63, https://www.ncbi.nlm.nih.gov/pubmed/29266851

20. K.C. Fettich and E.Y. Chen, 'Coping with obesity stigma affects depressed mood in African-American and white candidates for bariatric surgery', *Obesity* (Silver Spring), May 2012, 20(5), pp. 1118–21, https://www.ncbi.nlm.nih.gov/pubmed/22282108

21. S.E. Jackson, R.J. Beeken and J. Wardle, 'Perceived weight discrimination and changes in weight, waist circumference, and weight status', *Obesity* (Silver Spring), December 2014, 22(12), pp. 2485–8, https://www.ncbi.nlm.nih.gov/pubmed/25212272

22. B. Major, J.M. Hunger, D.P. Bunyan and C.T. Miller, 'The ironic effects of weight stigma', *Journal of Experimental Social Psychology* 74–80, https://www.sciencedirect.com/science/article/pii/S0022103113002047

6 HOW MUCH DOES OBESITY COST?

1. P. Scarborough, P. Bhatnagar, K.K. Wickramasinghe, S. Allender, C. Foster and M. Rayner. 'The economic burden of ill health due to diet, physical inactivity, smoking, alcohol and obesity in the UK: an update to 2006–07 NHS costs', *Journal of Public Health*, 1 December 2011, Volume 33, Issue 4, pp. 527–35, https://academic.oup.com/jpubhealth/article/33/4/527/1568587

2. B. Kennelly, 'How should cost-of-illness studies be interpreted?', *The Lancet*, October 2017, Volume 4, No. 10, pp. 735–6, https://www.thelancet.com/journals/lanpsy/article/PIIS2215-0366(17)30364-4/fulltext?code=lancet-sit

3. L.G.A. Bonneux, J.J. Barendregt, W.J. Nusselder and P.J. Van der Maas, 'Preventing fatal diseases increases healthcare costs: cause elimination life table approach', *British Journal of Medicine*, 1998, 316, p. 26, https://www.bmj.com/content/316/7124/26

4. Kennelly, 'How should cost-of-illness studies be interpreted?', pp. 735–6

5. P. Scarborough, P. Bhatnagar, K.K. Wickramasinghe, S. Allender, C. Foster and M. Rayner, Bapen Report, 'The cost of malnutrition in England and potential cost savings from nutritional interventions', 2016, https://www.bapen.org.uk/resources-and-education/publications-and-reports/malnutrition/cost-of-malnutrition-in-england

7 WHY ARE WE SO AFRAID OF FAT?

1. Ad Age Website, 'PETA tries fat shaming ad to convince people to go vegan', 17 September 2014, http://creativity-online.com/work/peta-go-vegan-bus-ad/37221

2. U. Irfan, 'Global Shift to Obesity Packs Serious Climate Consequences', *Scientific American*, June 2012, https://www.scientificamerican.com/article/global-shift-obesity-packs-serious-climate-consequences/

3. J.A. Swift, Hanlon, L. El-Redy, R.M. Puhl and C. Glazebrook, 'Weight bias among UK trainee dietitians, doctors, nurses and nutritionists', *Journal of Human Nutrition and Dietetics*, November 2012, https://www.ncbi.nlm.nih.gov/pubmed/23171227

4. S. Flint, 'Obesity stigma: Prevalence and impact in healthcare', *British Journal of Obesity*, 2015, Volume 1, No. 1, pp. 1–40, http://www.britishjournalofobesity.co.uk/journal/2015-1-1-14

5. S. Flint et al., 'Obesity Discrimination in the Recruitment Process: "You're Hired!"', *Frontiers in Psychology*, 2016, 7:647, https://www.frontiersin.org/articles/10.3389/fpsyg.2016.00647/full

6. D. E. White II, C. B. Wott and R. A. Carels, 'The Influence of Plaintiff's Body Weight on Judgments of Responsibility: The Role of Weight Bias', *Obesity Research and Clinical Practice*, November–December 2014, Volume 8, Issue 6,pp. e599–e607, https://www.sciencedirect.com/science/article/pii/S1871403X13002044

7. N.A. Schvey,Puhl, K.A. Levandoski andBrownell, 'The influence of a defendant's body weight on perceptions of guilt', *International Journal of Obesity* (London), September 2013, 37(9), pp. 1275–81, doi: 10.1038/ijo.2012.211, Epub 8 January 2013, https://www.ncbi.nlm.nih.gov/pubmed/23295503

8. Pascal's Pensées Blog, 'Is there a relationship between weight and success in PhD programs?', posted by Lascap on 8 June 2013, http://pensees.pascallisch.net/?p=1462

9. J.D. Latner, O'Brien, L.E. Durso, L.A. Brinkman, T. MacDonald, 'Weighing obesity stigma: the relative strength of different forms of bias', *International Journal of Obesity* (London), July 2008. 32(7), pp. 1145–52

10. L.A. Klos, C. Greenleaf, N. Paly, M.M. Kessler, C.G. Shoemaker, E.A. Suchla, 'Losing Weight on Reality TV: A Content Analysis of the Weight Loss Behaviors and Practices Portrayed on *The Biggest Loser*', *Journal of Health Communication*, 2015, 20(6), pp. 639–46, https://www.ncbi.nlm.nih.gov/pubmed/25909247

11. R. Cosslett and H. Baxter, *New Statesman*, 'Model scouts outside anorexia clinics highlight fashion's own don't ask, don't tell policy', 24 April 2013, https://www.newstatesman.com/media/2013/04/model-scouts-outside-anorexia-clinics-highlight-fashions-own-dont-ask-dont-tell-policy

12. S. Illing, 'Proof that Americans are lying about their sexual desires', Vox, 2 January 2018, https://www.vox.com/conversations/2017/6/27/15873072/google-porn-addiction-america-everybody-lies

13. D. McPhail, 'What to do with the "Tubby Hubby"? "Obesity," the Crisis of Masculinity, and the Nuclear Family in Early Cold War Canada', *Antipode*, November 2009, pp. 1021–50, https://onlinelibrary.wiley.com/doi/abs/10.1111/j.1467-8330.2009.00708.x

14. S. Flint, 'Obesity stigma: Prevalence and impact in healthcare', pp. 1–40

15. J.A. Swift, V. Tischler, S. Markham, I. Gunning, C. Glazebrook, C. Beer and R. Puhl, 'Are Anti-Stigma Films a Useful Strategy for Reducing Weight Bias Among Trainee Healthcare Professionals? Results of a Pilot Randomized Control Trial', *Obesity Facts*, March 2013, 6(1), pp. 91–102, https://www.ncbi.nlm.nih.gov/pmc/articles/PMC5644731/

PART II: WHY ARE WE SO FAT?
8 IS IT BECAUSE OF OUR GENES?

1. J. Friedman, 'Modern Science Versus the Stigma of Obesity', *Nature Medicine*, 2004,10,563–9,https://www.nature.com/articles/nm0604-563
2. C. Llewellyn and J. Wardle, 'Behavioral susceptibility to obesity: Gene-environment interplay in the development of weight', *Physiology and Behaviour*, 1 December 2015, 152(Pt B), pp. 494–501 https://www.ncbi.nlm.nih.gov/pubmed/26166156
3. N. Långström, Q. Rahman, E. Carlström and P. Lichtenstein, 'Genetic and Environmental Effects on Same-sex Sexual Behavior: A Population Study of Twins in Sweden', *Archives of Sexual Behaviour*, February 2010, Volume 39, Issue 1, pp. 75–80, https://link.springer.com/article/10.1007%2Fs10508-008-9386-1
4. Friedman, 'Modern Science Versus the Stigma of Obesity', pp. 563–9
5. Speakman et al., 'Set points, settling points and some alternative models', pp. 733–45
6. S. Cassels, 'Overweight in the Pacific: links between foreign dependence, global food trade, and obesity in the Federated States of Micronesia', *Globalisation and Health*, 2006, 2, p. 10, https://www.ncbi.nlm.nih.gov/pmc/articles/PMC1533815/
7. J.A. Hackett, R. Sengupta, J.J. Zylicz, K. Murakami, C. Lee, T.A. Down and M.A. Surani, 'Germline DNA demethylation dynamics and imprint erasure through 5-hydroxymethylcytosine', *Science*, 25 January 2013, 339(6118), pp. 448–52, https://www.ncbi.nlm.nih.gov/pubmed/23223451?dopt=Abstract&holding=npg
8 E. Watters, 'DNA Is Not Destiny: The New Science of Epigenetics', *Discovery*, November 2006, http://discovermagazine.com/2006/nov/cover
9. G.P. Ravelli, Z.A. Stein and M.W. Susser, 'Obesity in young men after famine exposure in utero and early infancy', *New England Journal of Medicine*, 12 August 1976, 295(7), pp. 349–53, https://www.ncbi.nlm.nih.gov/pubmed/934222?dopt=Abstract&holding=npg
10. B.T. Heijmans et al., 'Persistent epigenetic differences associated with prenatal exposure to famine in humans', *Proceedings of the National Academy of Science USA*, 4 November 2008, 105(44), pp. 17046–9, https://www.ncbi.nlm.nih.gov/pubmed/18955703?dopt=Abstract&holding=npg
11. S.F. Ng et al. 'Chronic high-fat diet in fathers programs β-cell dysfunction in female rat offspring', *Nature*, 21 October 2010, 467(7318), pp. 963–6, https://www.ncbi.nlm.nih.gov/pubmed/20962845?dopt=Abstract&holding=npg
12. Hebebrand et al., 'Epidemic obesity: are genetic factors involved via increased rates of assortative mating?', *International Journal of Obesity Related Metabolic Disorders*, March 2000, 24(3), pp. 345–53, https://www.ncbi.nlm.nih.gov/pubmed/10757629; Ajslev et al., 'Assortative marriages by body mass index have increased simultaneously with the obesity epidemic', *Frontiers in Genetics*, 18 July 2012, 3, pp. 125, https://www.ncbi.nlm.nih.gov/pubmed/23056005; Bastian et al., 'Number of children and the risk of obesity in older women', *Preventative Medicine*, January 2005, 40(1), pp. 99–104, https://www.ncbi.nlm.nih.gov/pubmed/15530586
13. Dawson et al., 'Propagation of obesity across generations: the roles of differential realized fertility and assortative mating by body mass index', *Human Heredity*, 2013, 75(2-4), pp. 204–12, https://www.ncbi.nlm.nih.gov/pubmed/24081235
14. S.M. Garn, T.V. Sullivan, V.M. Hawthorne, 'Educational level, fatness, and fatness differences between husbands and wives', *American Journal of Clinical Nutrition*,

October 1989, 50(4), pp. 740–45, https://www.ncbi.nlm.nih.gov/pubmed/2801578

15. P. Jacobson, J.S. Torgerson, L. Sjöström and C. Bouchard, 'Spouse resemblance in body mass index: effects on adult obesity prevalence in the offspring generation', *American Journal of Epidemiology*, 1 January 2007, 165(1), pp. 01–8, https://www.ncbi.nlm.nih.gov/pubmed/17041131

16. J.R. Speakman, K. Djafarian, J. Stewart and D.M. Jackson, 'Assortative mating for obesity', *American Journal of Clinical Nutrition*, August 2007, 86(2), pp. 316–23, https://www.ncbi.nlm.nih.gov/pubmed/17684200

17. https://www.psychologicalscience.org/observer/the-obesity-epidemic

18. J. Friedman, 'Obesity is Genetic', *Newsweek*, 9 September 2009, http://www.newsweek.com/obesity-genetic-79383

9 IS IT BECAUSE OF OUR GUTS?

1. A. Abbott, 'Scientists bust myth that our bodies have more bacteria than human cells', *Nature News*, 8 January 2016, https://www.nature.com/news/scientists-bust-myth-that-our-bodies-have-more-bacteria-than-human-cells-1.19136

2. Qin et al., 'A human gut microbial gene catalogue established by metagenomic sequencing', *Nature*,464,59–65,https://www.nature.com/articles/nature08821

3. Ibid.

4. C. Graham, A. Mullen and K. Whelan, 'Obesity and the gastrointestinal microbiota: a review of associations and mechanisms', *Nutrition Reviews*, 1 June 2015, Volume 73, Issue 6, pp. 376–85, https://academic.oup.com/nutritionreviews/article-abstract/73/6/376/1845882

5. Le Chatelier et al., 'Richness of human gut microbiome correlates with metabolic markers', *Nature*,29 August 2013, V500,541–6, https://www.nature.com/articles/nature12506

6. Ridaura et al., 'Cultured gut microbiota from twins discordant for obesity modulate adiposity and metabolic phenotypes in mice', *Science*https://www.ncbi.nlm.nih.gov/pmc/articles/PMC3829625/

7. N. Alang and C.R. Kelly, 'Weight Gain After Fecal Microbiota Transplantation', *Open Forum Infectious Diseases*, 1 January 2015, Volume 2, Issue 1, ofv004, https://academic.oup.com/ofid/article/2/1/ofv004/1461242

8. Vrieze et al., 'Transfer of intestinal microbiota from lean donors increases insulin sensitivity in individuals with metabolic syndrome', *Gastroenterology*, 2012, 143(4), pp. 913–16.e7, https://www.ncbi.nlm.nih.gov/pubmed/22728514

9. Chassaing et al., 'Dietary emulsifiers impact the mouse gut microbiota promoting colitis and metabolic syndrome', *Nature*,2015, 519(7541), pp. 92–6, https://www.nature.com/articles/nature14232

10. M.A. Sze and P.D. Schloss, 'Looking for a Signal in the Noise: Revisiting Obesity and the Microbiome', *mBio*, http://mbio.asm.org/content/7/4/e01018-16.abstract

11. Cotillard et al., 'Dietary intervention impact on gut microbial gene richness', *Nature*, 29 August 2013, 500,585–8, https://www.nature.com/articles/nature12480

10 IS IT BECAUSE OF CALORIES?

1. T.A. Smith, B.H. Lin and J.Y. Lee, 'Taxing Caloric Sweetened Beverages – Potential Effects on Beverage Consumption, Calorie Intake, and Obesity', USDA

Report, 2010, https://ageconsearch.umn.edu/bitstream/95465/2/ERR100.pdf; E.A. Finkelstein, C. Zhen, J. Nonnemaker and J.E. Todd, 'Impact of targeted beverage taxes on higher- and lower-income households', *Archives of Internal Medicine*, 13 December 2010, 170(22), pp. 2028–34, https://www.ncbi.nlm.nih.gov/pubmed/21149762; G.W. Gustavsen and K. Rickertsen, '*Applied Economics*Volume 43, Issue 6, pp. 707–16, https://www.tandfonline.com/doi/abs/10.1080/00036840802599776

2. Hall et al., 'Quantification of the effect of energy imbalance on bodyweight', *The Lancet*, https://www.ncbi.nlm.nih.gov/pmc/articles/PMC3880593/

3. https://www.niddk.nih.gov/health-information/weight-management/body-weight-planner

4. K.D. Hall, J. Guo, M. Dore and C.C. Chow, 'The Progressive Increase of Food Waste in America and Its Environmental Impact', *PLOS One*, 25 November 2009, http://journals.plos.org/plosone/article?id=10.1371/journal.pone.0007940

5. UK Government Family Food Statistics, https://www.gov.uk/government/collections/family-food-statistics

6. UK Government Diet and Nutrition Survey, https://www.gov.uk/government/collections/national-diet-and-nutrition-survey

7. R. Foster and J. Lunn, 'Food availability and our changing diet', British Nutrition Foundation Briefing Paper, 2007, https://www.nutrition.org.uk/attachments/144_Food%20availability%20and%20our%20changing%20diet.pdf

8. A.M. Prentice and S.A. Jebb, 'Obesity in Britain: gluttony or sloth?', pp. 437–9

9. Hollands et al., 'Portion, package or tableware size for changing selection and consumption of food, alcohol and tobacco', Cochrane Database of Systematic Reviews, 14 September 2015, http://www.cochrane.org/CD011045/PUBHLTH_portion-package-or-tableware-size-changing-selection-and-consumption-food-alcohol-and-tobacco

10. T. Bartlet, 'Spoiled Science – How a seemingly innocent blog post led to serious doubts about Cornell's famous food laboratory', *The Chronicle of Higher Education*, 17 March 2017, https://www.chronicle.com/article/Spoiled-Science/ 239529

11. T. van der Zee, J. Anaya and N.J.L. Brown, 'Statistical heartburn: An attempt to digest four pizza publications from the Cornell Food and Brand Lab', PeerJ Preprints, 25 January 2017, 5:e2748v1, https://peerj.com/preprints/2748/#aff-3

12. N. Brown, 'Some instances of apparent duplicate publication from the Cornell Food and Brand Lab', 2 March 2017, http://steamtraen.blogspot.com/2017/03/some-instances-of-apparent-duplicate.html

13. Statement from Brian Wansink, https://foodpsychology.cornell.edu/research-statement-april-2017

14. E. Robinson, S. Nolan, C. Tudur-Smith, E.J. Boyland, J.A. Harrold, C.A. Hardman and J.C.G. Halford, 'Will smaller plates lead to smaller waists? A systematic review and meta-analysis of the effect that experimental manipulation of dishware size has on energy consumption', *Obesity Reviews*, October 2014, https://onlinelibrary.wiley.com/doi/abs/10.1111/obr.12200

15. *Food Quality and Preference*, September 2018, Volume 68, pp. 80–89, https://www.sciencedirect.com/science/article/abs/pii/S0950329318301204#!

16. K. McCrickerd, L. Chambers, J.M. Brunstrom and M.R. Yeomans, 'Subtle changes in the flavour and texture of a drink enhance expectations of satiety', *Flavour*, https://flavourjournal.biomedcentral.com/articles/10.1186/2044-7248-1-20

17. K. McCrickerd and C.G. Forde, 'Sensory influences on food intake control:

moving beyond palatability', *Obesity Reviews*, January 2016, 17(1), pp. 18–29, https://www.ncbi.nlm.nih.gov/pubmed/26662879

18. K. McCrickerd, C.M. Lim, C. Leong, E.M. Chia and C.G. Forde, 'Texture-Based Differences in Eating Rate Reduce the Impact of Increased Energy Density and Large Portions on Meal Size in Adults', *The Journal of Nutrition*, June 2017, 147(6), pp. 1208–17, https://www.ncbi.nlm.nih.gov/pubmed/28446630; D.P. Bolhuis, C.G. Forde, Y. Cheng, H. Xu, N. Martin and C. de Graaf, 'Slow Food: Sustained Impact of Harder Foods on the Reduction in Energy Intake over the Course of the Day', *PLOS One*, 2 April 2014, http://journals.plos.org/plosone/article?id=10.1371/journal.pone.0093370

19. Benton and Young, 'Reducing Calorie Intake May Not Help You Lose Body Weight', pp. 703–14

20. C.G. Forde, N. van Kuijk, T. Thaler, C. de Graaf and N. Martin, 'Texture and savoury taste influences on food intake in a realistic hot lunch time meal', *Appetite*, January 2013, 60(1), pp. 180–86, https://www.ncbi.nlm.nih.gov/pubmed/23085683; McCrickerd et al., 'Texture-Based Differences in Eating Rate', pp. 1208–17

21. Ferriday et al., 'Variation in the Oral Processing of Everyday Meals Is Associated with Fullness and Meal Size; A Potential Nudge to Reduce Energy Intake?', *Nutrients*, 21 May 2016, 8(5), p. ii: e315, https://www.ncbi.nlm.nih.gov/pubmed/27213451

22. S.L. Tey, C.G. Forde, 'Impact of dose-response calorie reduction or supplementation of a covertly manipulated lunchtime meal on energy compensation', *Physiology and Behavior*, 15 October 2016, 165, pp. 15–21, https://www.ncbi.nlm.nih.gov/pubmed/27373874

23. D.M. Mourao, J. Bressan, W.W. Campbell Mattes, 'Effects of food form on appetite and energy intake in lean and obese young adults', *International Journal of Obesity*, 2007, 31,1688–95, https://www.nature.com/articles/0803667

11 IS IT BECAUSE WE ARE LAZY?

1. 'Patterns and Trends in Adult Obesity', Public Health England Presentation, 2017, https://www.slideshare.net/PublicHealthEngland/patterns-and-trends-in-adult-obesity

2. Casazza et al., 'Weighing the Evidence of Common Beliefs in Obesity Research', *Critical Reviews in Food Science and Nutrition*, 6 December 2016, https://www.ncbi.nlm.nih.gov/pmc/articles/PMC4272668/

3. H. Islam, L. K. Townsend, G.L. McKie, P.J. Medeiros, B.J. Gurd and T.J. Hazell, 'Potential involvement of lactate and interleukin-6 in the appetite-regulatory hormonal response to an acute exercise bout', *Journal of Applied Physiology*, 1 September 2017, 123(3), pp. 614–23, https://www.ncbi.nlm.nih.gov/pmc/articles/PMC5625078/

4. J. Friedman et al., 'Exercise training down-regulates ob gene expression in the genetically obese SHHF/Mcc-fa(cp) rat', *Hormone and Metabolic Research*, May 1997, 29(5), pp. 214–19, https://www.ncbi.nlm.nih.gov/pubmed/9228205

5. K. Westerterp, 'Physical activity, food intake, and body weight regulation: insights from doubly labelled water studies', *Nutrition Reviews*, 1 March 2010, Volume 68, Issue 3, pp. 148–54, https://academic.oup.com/nutritionreviews/article-abstract/68/3/148/1910457?redirectedFrom=fulltext

6. Ibid.

7. A. Malhotra, T. Noakes, S. Phinney, 'It is time to bust the myth of physical inactivity and obesity: you cannot outrun a bad diet', *British Journal of Sports Medicine*, August 2015, 49(15), pp. 967–8, https://www.ncbi.nlm.nih.gov/pubmed/25904145

8. Kerns et al., 'Increased Physical Activity Associated with Less Weight Regain Six Years After "The Biggest Loser" Competition', *Obesity* (Silver Spring), November 2017, 25(11), pp. 1838–43, https://www.ncbi.nlm.nih.gov/pubmed/29086499

9. D.L. Swift, N.M. Johannsen, C.J. Lavie, C.P. Earnest and T.S. Church, 'The Role of Exercise and Physical Activity in Weight Loss and Maintenance', *Progress in Cardiovascular Diseases*, 2014, 56(4), pp. 441–7, https://www.ncbi.nlm.nih.gov/pmc/articles/PMC3925973/

10. M.E.J. Lean and D. Malkova, 'Altered gut and adipose tissue hormones in overweight and obese individuals: cause or consequence?', *International Journal of Obesity*, April 2016, 40(4), pp. 622–32, https://www.ncbi.nlm.nih.gov/pubmed/26499438

11. E. Zschucke, K. Gaudlitz and A. Ströhle, 'Exercise and Physical Activity in Mental Disorders: Clinical and Experimental Evidence', *Journal of Preventative Medicine and Public Health*, January 2013, 46(Suppl 1), pp. S12–S21, https://www.ncbi.nlm.nih.gov/pmc/articles/PMC3567313/

12. R.J.H.M. Verheggen, M.F.H. Maessen, D.J. Green, A.R.M.M. Hermus, M.T.E. Hopman and D.H.T. Thijssen, 'A systematic review and meta-analysis on the effects of exercise training versus hypocaloric diet: distinct effects on body weight and visceral adipose tissue', *Obesity Reviews*, https://onlinelibrary.wiley.com/doi/abs/10.1111/obr.12406

13. K. Do, R.E. Brown, S. Wharton, C.I. Ardern and J.L. Kuk, 'Association between cardiorespiratory fitness and metabolic risk factors in a population with mild to severe obesity', *BMC Obesity*, 2018, 5:5, https://bmcobes.biomedcentral.com/track/pdf/10.1186/s40608-018-0183-7?site=bmcobes.biomedcentral.com

14. J.H. Wilmore, J.P. Després, P.R. Stanforth, S. Mandel, T. Rice, J. Gagnon, A. S. Leon, D.C. Rao, J.S. Skinner and C. Bouchard, 'Alterations in body weight and composition consequent to 20 weeks of endurance training: the HERITAGE Family Study', *American Journal of Clinical Nutrition*, 1999, 70, pp. 346–52, https://pdfs.semanticscholar.org/6399/1ce2c6f7d9243c6de1f87c61b1e4d d1d6574.pdf

15. P. Ekkekakis, S. Vazou, W.R. Bixby, E. Georgiadis, 'The mysterious case of the public health guideline that is (almost) entirely ignored: call for a research agenda on the causes of the extreme avoidance of physical activity in obesity', *Obesity Reviews*, April 2016, Volume 17, Issue 4, pp. 313–29, https://onlinelibrary.wiley.com/doi/full/10.1111/obr.12369

16. A. Myers, C. Gibbons, G. Finlayson and J. Blundell, 'Associations among sedentary and active behaviours, body fat and appetite dysregulation: investigating the myth of physical inactivity and obesity', *British Journal of Sports Medicine*, 4 April 2016, http://bjsm.bmj.com/content/51/21/1540?ct

17. S.W. Flint and S. Reale, 'Weight stigma in frequent exercisers: Overt, demeaning and condescending', *Journal of Health Psychology*, 4 July 2016, http://journals.sagepub.com/doi/abs/10.1177/1359105316656232

12 IS IT BECAUSE OF FAT?

1. C.A. Running, B.A. Craig and R.D. Mattes, 'Oleogustus: The Unique Taste of Fat', *Chemical Senses*, 1 September 2015, Volume 40, Issue 7, pp. 507–16, https://academic.oup.com/chemse/article/40/7/507/400784

2. K. Hall, 'Did the Food Environment Cause the Obesity Epidemic?', *Obesity – A Research Journal*, 20 December 2017, Volume 26, Issue 1, https://onlinelibrary. wiley.com/doi/full/10.1002/oby.22073; J.P. Flatt, 'Use and storage of carbohydrate and fat', *American Journal of Clinical Nutrition*, April 1995, 61(4 Suppl), pp. 952S–9S, https://www.ncbi.nlm.nih.gov/pubmed/7900694

3. G.A. Bray and B.M. Popkin, 'Dietary fat intake does affect obesity!', *American Journal of Clinical Nutrition*, December 1998, 68(6), pp. 1157–73, https://www. ncbi.nlm.nih.gov/pubmed/9846842

4. Trends in Coronary Heart Disease 1961–2011, British Heart Foundation Report, 2011, https://www.bhf.org.uk/publications/statistics/trends-in-coronary-heart-disease-1961-2011

5. L. Hooper, N. Martin, A. Abdelhamid and G.D. Smith, 'Reduction in saturated fat intake for cardiovascular disease', Cochrane Database of Systematic Reviews, 10 June 2015, http://cochranelibrary-wiley.com/doi/10.1002/14651858.CD011737/abstract

6. W.C. Willett and A. Ascherio, 'Trans fatty acids: are the effects only marginal?', *American Journal of Public Health*, 1 May 1994, 84, No. 5, pp. 722–4, https:// ajph.aphapublications.org/doi/10.2105/AJPH.84.5.722

7. Ascherio et al., '1999, 340, pp. 1994–8, https://www.nejm.org/doi/full/10.1056/ nejm199906243402511

8. No authors listed, 'Trans fatty acids and coronary heart disease risk', Report of the expert panel on trans fatty acids and coronary heart disease, *American Journal of Clinical Nutrition*, September 1995, 62(3), https://www.ncbi.nlm.nih.gov/ pubmed/7661131

9. M.P. Iqbal, 'Trans fatty acids – A risk factor for cardiovascular disease', *Pakistan Journal of Medical Sciences*, January–February 2014, 30(1), pp. 194–7, https:// www.ncbi.nlm.nih.gov/pmc/articles/PMC3955571/

10. M.S. Butt and M.T. Sultan, 'Levels of trans fats in diets consumed in developing economies', *J AOAC Int.*,2009, 92(5), https://www.ncbi.nlm.nih.gov/pubmed/ 19916365

11. P.J. Jones, 'Dietary cholesterol and the risk of cardiovascular disease in patients: a review of the Harvard Egg Study and other data', *International Journal of Clinical Practice,*https://www.researchgate.net/publication/26809478_Dietary_ cholesterol_and_the_risk_of_cardiovascular_disease_in_patients_A_review_of_ the_Harvard_Egg_Study_and_other_data

12. Hu et al., 'A prospective study of egg consumption and risk of cardiovascular disease in men and women', *JAMA*, 21 April 1999, 281(15), pp. 1387–94, https://www.ncbi.nlm.nih.gov/pubmed/10217054

13. D. Mozaffarian, R. Micha and S. Wallace, 'Effects on Coronary Heart Disease of Increasing Polyunsaturated Fat in Place of Saturated Fat: A Systematic Review and Meta-Analysis of Randomized Controlled Trials', *PLOS Medicine*, 23 March 2010, http://journals.plos.org/plosmedicine/article?id=10.1371/journal.pmed.1000252

14. Mozaffarian et al., 'Interplay between different polyunsaturated fatty acids and risk of coronary heart disease in men', *Circulation*, 18 January2005, 111(2), pp. 157–64, https://www.ncbi.nlm.nih.gov/pubmed/15630029/

15. A. Ascherio, Rimm, E.L. Giovannucci, D. Spiegelman, M. Stampfer and W.C. Willett, 'Dietary fat and risk of coronary heart disease in men: cohort follow up study in the United States', *BMJ*,1996, 313(7049), pp. 84–90, https://www.ncbi. nlm.nih.gov/pubmed/8688759/

16. P. Leren, 'The Oslo Diet-Heart Study – Eleven-Year Report', *Circulation*, http:// circ.ahajournals.org/content/42/5/935

17. Wang et al., 'Association of Specific Dietary Fats With Total and Cause-Specific Mortality', *JAMA Internal Medicine*, 1 August 2016, 176(8), pp. 1134–45, https://www.ncbi.nlm.nih.gov/pubmed/27379574

18. T.A. Lennie, M.L. Chung, D.L. Habash and D.K.Moser, 'Dietary fat intake and proinflammatory cytokine levels in patients with heart failure', *Journal of Cardiac Failure*, October 2005, 11(8), pp. 613–18, https://www.ncbi.nlm.nih.gov/pubmed/ 16230265/

19. Griffin et al., 'Effects of altering the ratio of dietary n-6 to n-3 fatty acids on insulin sensitivity, lipoprotein size, and postprandial lipemia in men and post-menopausal women aged 45–70 y: the OPTILIP Study', *American Journal of Clinical Nutrition*, December 2006, 84(6), pp. 1290–98, https://www.ncbi.nlm. nih.gov/pubmed/17158408

20. Expert reaction to editorial in *Open Heart* journal on omega 6 and omega 3 fatty acids and obesity, 24 October 2016, http://www.sciencemediacentre.org/ expert-reaction-to-editorial-on-omega-6-and-omega-3-fatty-acids-and-obesity/

21. Stanley et al., 'UK Food Standards Agency Workshop Report: the effects of the dietary n-6:n-3 fatty acid ratio on cardiovascular health', *British Journal of Nutrition*, December 2007, 98(6), pp. 1305–10, https://www.ncbi.nlm.nih.gov/ pubmed/18039412

22. Joint FAO/WHO Expert Consultation on Fats and Fatty Acids in Human Nutrition, 10–14 November 2008, WHO, Geneva: Interim Summary of Conclusions and Dietary Recommendations on Total Fat & Fatty Acids http:// www.who.int/nutrition/topics/FFA_summary_rec_conclusion.pdf

13 IS IT BECAUSE OF CARBS?

1. W. Kaufman, 'Atkins Bankruptcy a Boon for Pasta Makers', 3 August 2005, https://www.npr.org/templates/story/story.php?storyId=4783324

2. Pliquett et al., 'The effects of insulin on the central nervous system—focus on appetite regulation', *Hormone and Metabolic Research*, July 2006, 38(7), pp. 442–6, https://www.ncbi.nlm.nih.gov/pubmed/16933179?itool=EntrezSystem2. PEntrez.Pubmed.Pubmed_ResultsPanel.Pubmed_RVDocSum&ordinalpos=66

3. J.D. Roth, H. Hughes. E. Kendall, A.D. Baron and C.M. Anderson, 'Antiobesity effects of the beta-cell hormone amylin in diet-induced obese rats: effects on food intake, body weight, composition, energy expenditure, and gene expression', *Endocrinology*, December 2006, 147(12), pp. 5855–64, https://www.ncbi.nlm. nih.gov/pubmed/16935845

4. D.A.Vanderweele, F.X. Pi-Sunyer, D. Novin, M.J. Bush, 'Chronic insulin infusion suppresses food ingestion and body weight gain in rats', *Brain Research Bulletin*, Volume 5, Supplement 4,pp. 7–11, https://www.sciencedirect.com/ science/article/abs/pii/0361923080902233#!

5. M.D. Jensen, M.W. Haymond, R.A. Rizza, P.E. Cryer and J.M. Miles, 'Influence of body fat distribution on free fatty acid metabolism in obesity', *The Journal of*

Clinical Investigation, April 1989, 83(4), pp. 1168–73, https://www.ncbi.nlm.
nih.gov/pmc/articles/PMC303803/

6. Makimura et al., 'Metabolic Effects of Long-Term Reduction in Free Fatty Acids
   With Acipimox in Obesity: A Randomized Trial', *The Journal of Clinical
   Endocrinology and Metabolism*, 21 December 2015, http://press.endocrine.org/
   doi/10.1210/jc.2015-3696
7. E. Boelsma, E.J. Brink, A. Stafleu, H.F. Hendriks, 'Measures of postprandial
   wellness after single intake of two protein-carbohydrate meals', *Appetite*, June
   2010, 54(3), pp. 456–64, https://www.ncbi.nlm.nih.gov/pubmed/20060863
8. Tentolouris et al., 'Diet-induced thermogenesis and substrate oxidation are not
   different between lean and obese women after two different isocaloric meals,
   one rich in protein and one rich in fat', *Metabolism*,2008, 57(3), pp.313–20,
   https://www.ncbi.nlm.nih.gov/pubmed/18249201
9. C. Weyer, Bogardus, R.E. Pratley, 'Metabolic factors contributing to increased
   resting metabolic rate and decreased insulin-induced thermogenesis during the
   development of type 2 diabetes', *Diabetes*,1999, 48(8), pp. 1607–14, https://
   www.ncbi.nlm.nih.gov/pubmed/10426380
10. Hill et al., 'Nutrient balance in humans: effects of diet composition', *American
    Journal of Clinical Nutrition*, July 1991, 54(1), pp. 10–7, https://www.ncbi.nlm.
    nih.gov/pubmed/2058571
11. J. Hirsch, L.C. Hudgins, R.L. Leibel, M. Rosenbaum, 'Diet composition and
    energy balance in humans', *American Journal of Clinical Nutrition*, March 1998,
    67(3 Suppl), pp. 551S–5S, https://www.ncbi.nlm.nih.gov/pubmed/9497169
12. R. Sichieri, Moura, V. Genelhu, F. Hu, W.C. Willett, 'An 18-mo randomized trial
    of a low-glycemic-index diet and weight change in Brazilian women', *American
    Journal of Clinical Nutrition*, September 2007, 86(3), pp. 707–13, https://www.
    ncbi.nlm.nih.gov/pubmed/17823436
13. S. O'Rahilly, 'Human genetics illuminates the paths to metabolic disease',
    Nature, 2009, 462(7271), pp. 307–14, https://www.ncbi.nlm.nih.gov/pubmed/
    19924209
14. I.S. Farooqi, 'Monogenic Human Obesity – Obesity and Metabolism', *Frontiers
    of Hormone Research*, Basel, Karger, 2008, Volume 36, pp. 1–11, https://www.
    ncbi.nlm.nih.gov/pubmed/18230891
15. Graph of US food intake prepared by Stephan Guyenet, http://1.bp.blogspot.
    com/-zmPvM2rFwUw/UFO0mUtZOFI/AAAAAAAABFA/iTgRqWO5zgs/
    s1600/Adjusted+macro+intake+1909-2006.jpg
16. I. Elmadfa and A.L. Meyer, 'The Debate Goes on: New Evidence for the Role of
    Macronutrient Distribution on Body Weight Development', *EBio Medicine*,
    October 2017, 24, pp. 32–3, https://www.ncbi.nlm.nih.gov/pmc/articles/
    PMC5652134/
17. Hall et al., 'Calorie for Calorie, Dietary Fat Restriction Results in More Body Fat
    Loss than Carbohydrate Restriction in People with Obesity', *Cell Metabolism*, 1
    September 2015, 22(3), pp. 427–36, https://www.ncbi.nlm.nih.gov/pubmed/
    26278052
18. Hall et al., 'Energy expenditure and body composition changes after an isocaloric
    ketogenic diet in overweight and obese men', *The American Journal of Clinical
    Nutrition*, 1 August 2016, Volume 104, Issue 2, https://academic.oup.com/ajcn/
    article/104/2/324/4564649

19. M.J. Dekker, Q. Su, C. Baker, A.C. Rutledge and K. Adeli, 'Fructose: a highly lipogenic nutrient implicated in insulin resistance, hepatic steatosis, and the metabolic syndrome', *American Journal of Physiology, Endocrinology and Metabolism*, November 2010, 299(5), pp. E685–94, https://www.ncbi.nlm.nih.gov/pubmed/20823452

20. J. Lambertz, T. Berger, T.W. Mak, J. van Helden and R. Weiskirchen, 'Lipocalin-2 in Fructose-Induced Fatty Liver Disease', *Frontiers in Physiology*, 28 November 2017, https://www.frontiersin.org/articles/10.3389/fphys.2017.00964/full

21. J. Steen, 'So, This Is Exactly How Sugar Makes Us Fat', 21 April 2017, https://www.huffingtonpost.com.au/2017/04/20/so-this-is-exactly-how-sugar-makes-us-fat_a_22046969/?guccounter=1

22. K.F. Petersen and G.I. Shulman, 'Etiology of insulin resistance', p. S10–16

23. SACN Report 2015 – Carbohydrates and Health, https://assets.publishing.service.gov.uk/government/uploads/system/uploads/attachment_data/file/445503/SACN_Carbohydrates_and_Health.pdf

14   IS IT BECAUSE WE ARE ADDICTED?

1. A. Meadows, L.J. Nolan and S. Higgs, 'Self-perceived food addiction: Prevalence, predictors, and prognosis', *Appetite*,2017, 114:282–98, https://www.ncbi.nlm.nih.gov/pubmed/28385581

2. N.M. Lee, J. Lucke, W.D. Hall, C. Meurk, F.M. Boyle and A. Carter, 'Public Views on Food Addiction and Obesity: Implications for Policy and Treatment', *PLOS One*, 25 September 2013, http://journals.plos.org/plosone/article?id=10.1371/journal.pone.0074836

3. H. Ziauddeen, I.S. Farooqi and P.C. Fletcher, 'Obesity and the brain: how convincing is the addiction model?', *Nature Reviews Neuroscience*,13,279–86, https://www.nature.com/articles/nrn3212

4. Ibid.

5. C.M. Grilo and R.M. Masheb, 'Childhood psychological, physical, and sexual maltreatment in outpatients with binge eating disorder: frequency and associations with gender, obesity, and eating-related psychopathology', *Obesity Research*, May 2001, 9(5), pp. 320–25, https://www.ncbi.nlm.nih.gov/pubmed/11346674

6. J.D. Latner, R.M. Puhl, J.M. Murakami and K.S. O'Brien, 'Food addiction as a causal model of obesity. Effects on stigma, blame, and perceived psychopathology', *Appetite*, 1 June 2014, Volume 77,pp. 79–84, https://www.sciencedirect.com/science/article/pii/S0195666314001196#!

15   IS IT BECAUSE OF OUR ENVIRONMENT?

1. Bartoshuk, The "Obesity Epidemic"'

2. R.C. Davey, 'The obesity epidemic: too much food for thought?', *British Journal of Sports Medicine*, 2004, 38, pp. 360–63, https://bjsm.bmj.com/content/38/3/360

3. UK Government Family Food Statistics, ; UK Government Diet and Nutrition Survey, https://www.gov.uk/government/collections/national-diet-and-nutrition-survey

4. K. Langlois, D. Garriguet, L.Findlay, 'Diet composition and obesity among Canadian adults', *Health Reports*, December 2009, 20(4), pp. 11–20, https://www.ncbi.nlm.nih.gov/pubmed/20108602

5. Casazza et al., 'Weighing the Evidence of Common Beliefs in Obesity Research',

6. E. Oken, E.B. Levitan, M.W. Gillman, 'Maternal smoking during pregnancy and child overweight: systematic review and meta-analysis', *International Journal of Obesity*, 32, 201–10, https://www.nature.com/articles/0803760

7. Y. Goryakin, T. Lobstein, W.P.T. James, M. Suhrcke, 'The impact of economic, political and social globalization on overweight and obesity in the 56 low and middle income countries', *Social Science and Medicine*, May 2015, 133, pp.67–76, https://www.ncbi.nlm.nih.gov/pmc/articles/PMC4416723/

8. K.D Hall, J. Guo, M. Dore and C.C. Chow, 'The Progressive Increase of Food Waste in America and Its Environmental Impact', *PLOS ONE*, November 2009, https://journals.plos.org/plosone/article?id=10.1371/journal.pone.0007940

9. Casazza et al., 'Weighing the Evidence of Common Beliefs in Obesity Research', *Critical Reviews in Food Science and Nutrition*, December 2015, 55(14), pp. 2014–2053, https://www.ncbi.nlm.nih.gov/pmc/articles/PMC4272668/

10. H. Lee, 'The role of local food availability in explaining obesity risk among young school-aged children', *Social Science and Medicine*, April 2012, 74(8), pp. 1193–203, https://www.ncbi.nlm.nih.gov/pubmed/22381683; R. An and R. Sturm, 'School and residential neighborhood food environment and diet among California youth', *American Journal of Preventative Medicine*, February 2012, 42(2), pp. 129–35, https://www.ncbi.nlm.nih.gov/pubmed/22261208

11. Food environments and obesity – neighbourhood or nation?', *International Journal of Epidemiology*, 1 February 2006, Volume 35, Issue 1, https://academic.oup.com/ije/article/35/1/100/849977

12. T. Pearson, J. Russell, M.J. Campbell and M.E. Barker, 'Do "food deserts" influence fruit and vegetable consumption? A cross-sectional study', *Appetite*, October 2005, 45(2), pp. 195–7, https://www.ncbi.nlm.nih.gov/pubmed/15927303

13. Food environments and obesity – neighbourhood or nation?, Ibid.

14. A.M. Prentice and S.A. Jebb, 'Fast foods, energy density and obesity: a possible mechanistic link', *Obesity Reviews*, https://onlinelibrary.wiley.com/doi/full/10.1046/j.1467-789X.2003.00117.x; B.J. Rolls, ' The Supersizing of America: Portion Size and the Obesity Epidemic', *Nutrition Today*, March-April 2003, Volume 38, Issue 2, pp. 42–53, https://journals.lww.com/nutritiontodayonline/Abstract/2003/03000/The_Supersizing_of_America__Portion_Size_and_the.4.aspx

15. M.A. Pereira, A.I. Kartashov, C.B. Ebbeling, L. Van Horn, M.L. Slattery, D.R. Jacobs Jr and D.S. Ludwig , 'Fast-food habits, weight gain, and insulin resistance (the CARDIA study): 15-year prospective analysis', *The Lancet*, 1–7 January 2005, 365(9453), pp. 36–42, https://www.ncbi.nlm.nih.gov/pubmed/15639678

16. Forouhi et al., 'Associations between exposure to takeaway food outlets, takeaway food consumption, and body weight in Cambridgeshire, UK: population based, cross sectional study', *BMJ*, https://www.bmj.com/content/348/bmj.g1464

17. D. Simmons, A. McKenzie, S. Eaton, N. Cox, M.A. Khan, J. Shaw, P. Zimmet, 'Choice and availability of takeaway and restaurant food is not related to the prevalence of adult obesity in rural communities in Australia', *International Journal of Obesity*, 2005, 29,703–10, https://www.nature.com/articles/0802941; H.L. Burdette and R.C. Whitaker, 'Neighborhood playgrounds, fast food restaurants, and crime: relationships to overweight in low-income preschool children', *Preventative Medicine*, January 2004, 38(1), pp. 57–63, https://www.ncbi.nlm.nih.gov/pubmed/14672642

18. A. Rodgers, A. Woodward, B. Swinburn and W.H. Dietz, 'Prevalence trends tell

us what did not precipitate the US obesity epidemic', *The Lancet Public Health*, April 2018, Volume 3, No. 4, pp. e162–3, https://www.thelancet.com/journals/lanpub/article/PIIS2468-2667(18)30021-5/fulltext

19. A. Kirkland, 'The environmental account of obesity: a case for feminist skepticism', *Signs* (Chicago), 2011, 36(2), pp. 463–86, https://www.ncbi.nlm.nih.gov/pubmed/21114084

20. Pellegrino et al., 'A Novel Variant is Associated with Short Sleep and Resistance to Sleep Deprivation in Humans', *Sleep*, 1 August 2014, 37(8), pp. 1327–36, https://www.ncbi.nlm.nih.gov/pmc/articles/PMC4096202/

21. Y.S. Bin, N.S. Marshall, N. Glozier, 'Secular trends in adult sleep duration: a systematic review' *Sleep Medicine Reviews*, June 2012, 16(3), pp. 223–0., https://www.ncbi.nlm.nih.gov/pubmed/?term=Secular+trends+in+adult+sleep+duratio n%3A+a+systematic+review

22. Yetish et al., 'Natural Sleep and Its Seasonal Variations in Three Pre-industrial Societies', *Current Biology*, 2 November 2015, Volume 25, Issue 21, pp. 2862–8, https://www.cell.com/current-biology/fulltext/S0960-9822(15)01157-4

23. Knutson et al., 'Trends in the prevalence of short sleepers in the USA: 1975–2006', *Sleep*, 2010, 33(1), pp. 37–45, https://www.ncbi.nlm.nih.gov/pubmed/20120619

24. K. Eckel-Mahan and P. Sassone-Cors, 'Metabolism and the Circadian Clock Converge', *Physiological Reviews*, 1 January 2013, https://www.physiology.org/doi/10.1152/physrev.00016.2012

25. D. Mozaffarian, T. Hao, E.B. Rimm, W.C. Willett and F.B. Hu, 'Changes in diet and lifestyle and long-term weight gain in women and men', *New England Journal of Medicine*, 23 June 2011, 364 (25), pp. 2392–404, https://www.ncbi.nlm.nih.gov/pubmed/21696306

26. Reilly et al., 'Early life risk factors for obesity in childhood: cohort study', *BMJ*, 11 June 2005, 330(7504), p. 1357, https://www.ncbi.nlm.nih.gov/pubmed/15908441

27. Owens et al., 'Television-viewing habits and disturbance in school children', *Pediatrics*, 1999, 104(3), p. e27, https://www.ncbi.nlm.nih.gov/pubmed/10469810

28. T. Roenneberg, K. V. Allebrandt, M. Merrow and C. Vetter, 'Social jetlag and obesity', *Current Biology*, 22 May 2012, 22(10), pp. 39–43, https://www.science-direct.com/science/article/pii/S0960982212003259

29. McAllister et al., 'Ten Putative Contributors to the Obesity Epidemic', *Critical Reviews in Food Science and Nutrition*, November https://www.ncbi.nlm.nih.gov/pmc/articles/PMC2932668/

30. B. Buemann, A. Astrup, N.J. Christensen and J. Madsen, 'Effect of moderate cold exposure on 24-h energy expenditure: similar response in postobese and nonobese women', *American Journal of Physiology*, December 1992, 263(6 Pt 1), pp. e1040–5, https://www.ncbi.nlm.nih.gov/pubmed/1476175/; M.S. Westerterp-Plantenga, W.D. van Marken Lichtenbelt, C. Cilissen and S. Top, 'Energy metabolism in women during short exposure to the thermoneutral zone', *Physiology and Behaviour*, 1–15 February 2002, 75(1–2), pp. 227–35, https://www.ncbi.nlm.nih.gov/pubmed/11890972/

31. E.A. Rowe, B.J. Rolls, 'Effects of environmental temperature on dietary obesity and growth in rats', *Physiology and Behaviour*, February 1982, 28(2), pp. 219–26, https://www.ncbi.nlm.nih.gov/pubmed/7079333/

32. McAllister et al. 'Ten Putative Contributors to the Obesity Epidemic', *Critical Reviews in Food Science and Nutrition* 2009 Nov; 49(10): 868–913. https://www.ncbi.nlm.nih.gov/pmc/articles/PMC2932668/

33. A. Vargas-Castillo, R. Fuentes-Romero, L.A. Rodriguez-Lopez, N. Torres and A.R. Tovar, 'Understanding the Biology of Thermogenic Fat: Is Browning A New Approach to the Treatment of Obesity?', *Archives of Medical Research*, July 2017, 48(5), pp. 401–13, https://www.ncbi.nlm.nih.gov/pubmed/29102386

34. S. Virtue and A. Vidal-Puig, 'It's not how fat you are, it's what you do with it that counts', *PLOS Biology*, 23 September 2008, 6(9), p. e237, http://journals.plos.org/plosbiology/article?id=10.1371/journal.pbio.0060237

35. McAllister et al. 'Ten Putative Contributors to the Obesity Epidemic', pp. 868–913; Patterson et al., 'Sociodemographic factors and obesity in preadolescent black and white girls: NHLBI's Growth and Health Study', *Journal of the National Medical Association*, September 1997, 89(9), pp. 594–600, https://www.ncbi.nlm.nih.gov/pubmed/9302856?dopt=Abstract&holding=npg; Keith et al., 'Putative contributors to the secular increase in obesity: exploring the roads less travelled', *International Journal of Obesity*, 2006, 30,1585–94 https://www.nature.com/articles/0803326

36. C. Pelletier, P. Imbeault and A. Tremblay, 'Energy balance and pollution by organochlorines and polychlorinated biphenyls', *Obesity Reviews*, February 2003, 4(1), pp. 17–24, https://www.ncbi.nlm.nih.gov/pubmed/12608524?dopt=Abstract&holding=npg; Keith et al. 'Putative contributors to the secular increase in obesity', pp. 1585–94; McAllister et al., 'Ten Putative Contributors to the Obesity Epidemic', pp. 868–913

16 IS IT BECAUSE WE ARE POOR?

1. Hill et al., 'Stress and eating behaviors in children and adolescents:and meta-analysis', *Appetite*, 1 April 2018, 123, pp. 14–22, https://www.ncbi.nlm.nih.gov/pubmed/29203444

2. R. Rosmond, 'Role of stress in the pathogenesis of the metabolic syndrome', *Psychoneuroendocrinology*, January 2005, 30(1), pp.1–10, https://www.ncbi.nlm.nih.gov/pubmed/15358437; P. Björntorp, 'Hormonal control of regional fat distribution', *Human Reproduction*, October 1997, 12 Suppl 1, pp. 21–5, https://www.ncbi.nlm.nih.gov/pubmed/9403318

3. M. Lieberman, *Social: Why Our Brains Are Wired to Connect*

4. S. Junger, *Tribe: On Homecoming and Belonging* (London: Fourth Estate, 2016)

5. John Bingham, 'Lonely Britain: five million people who have no real friends', *The Telegraph*, August 2014, https://www.telegraph.co.uk/news/uknews/11026520/Lonely-Britain-five-million-people-who-have-no-real-friends.html

6. Survey from 'Time To Change', reported 1 February 2018 on BBC website, 'Two thirds of UK adults have "nobody to talk to" about problems', http://www.bbc.co.uk/news/health-42903914

7. M. McPherson, L. Smith-Lovin and M.E. Brashears, 'Social Isolation in America: Changes in Core Discussion Networks over Two Decades', *American Sociological Review*, 1 June 2006, http://journals.sagepub.com/doi/abs/10.1177/000312240607100301

8. C. Dudley, J. Kerns and K. Steadman, Work Foundation Report – *Gender, Sex, Health and Work*, http://www.theworkfoundation.com/wp-content/uploads/2016/10/418_GSHWBackground.pdf

9. J. Raisborough, *Fat Bodies, Health and the Media* (London: Palgrave Macmillan, 2016)

PART III: WHAT SHOULD WE DO?
17 SURGERY AND DRUGS

1. Weintraub et al., 'Long-term weight control study. I (weeks 0 to 34). The enhancement of behavior modification, caloric restriction, and exercise by fenfluramine plus phentermine versus placebo', *Clinical Pharmacology and Therapeutics*, May 1992, 51(5), pp. 586–94, https://www.ncbi.nlm.nih.gov/pubmed/1587072

2. 'Cardiac Valvulopathy Associated with Exposure to Fenfluramine or Dexfenfluramine', US Department of Health and Human Services Interim Public Health, November 1997, 46(45), pp. 1061–6, https://www.cdc.gov/mmwr/preview/mmwrhtml/00049815.htm

3. F.A. Moreira and J.A. Crippa , 'The psychiatric side-effects of rimonabant', *Revista Brasileira de Psiquiatria*, June 2009, 31(2), pp. 145–53, https://www.ncbi.nlm.nih.gov/pubmed/19578688

4. D. Rucker, R. Padwal, S.K. Li, C. Curioni and D.C. Lau, 'Long term pharmaco-therapy for obesity and overweight: updated meta-analysis', *BMJ,*2007, 335(7631), pp. 1194–9, https://www.ncbi.nlm.nih.gov/pubmed/18006966

5. J.A. Madura II and J.K. DiBaise, 'Quick fix or long-term cure? Pros and cons of bariatric surgery', *F1000 Reports Medicine*, 2012, 4, p. 19, https://www.ncbi.nlm.nih.gov/pmc/articles/PMC3470459/; Angrisani et al. 'Bariatric Surgery Worldwide 2013', *Obesity Surgery*, October 2015, 25(10), pp. 1822–32, https://www.ncbi.nlm.nih.gov/pubmed/25835983

6. J.B. Dixon, E.A. Lambert and G.W. Lambert, 'Neuroendocrine adaptations to bariatric surgery', *Molecular and Cellular Endocrinology*, 418, pp. 143–52, http://doi.org/10.1016/j.mce.2015.05.033

7. Pareek et al., 'Metabolic Surgery', *Journal of the American College of Cardiology*, 2018, 71(6), pp. 670–87, http://doi.org/10.1'016/j.jacc.2017.12.014

8. S.A. Brethauer, B.Chand and P.R. Schauer, 'Risks and benefits of bariatric surgery: current evidence', *Cleveland Clinic Journal of Medicine*, November 2006, 73(11), pp. 993–1007, https://www.ncbi.nlm.nih.gov/pubmed/17128540

9. Buchwald et al, 'Bariatric surgery: a systematic review and meta-analysis',*JAMA*,2004, 292(14), pp. 1724–37, https://www.ncbi.nlm.nih.gov/pubmed/15479938

10. Spadola et al., 'Alcohol and drug use among post-operative bariatric patients: A systematic review of the emerging research and its implications', *Alcoholism Clinical and Experimental Research*, September, https://www.ncbi.nlm.nih.gov/pmc/articles/PMC4608681/

11. King et al., 'Alcohol and other substance use after bariatric surgery: prospective evidence from a U.S. multicenter cohort study', *Surgery for Obesity and Related Diseases*, 2017, Volume 13, Issue 8, pp. 1392–402, https://www.soard.org/article/S1550-7289(17)30152-1/fulltext

12. NICE Clinical Guidelines (CG189), November 2014, 'Obesity: identification, assessment and management', https://www.nice.org.uk/guidance/cg189/chapter/1-recommendations#surgical-interventions

13. R. Welbourn, C.W. le Roux, A. Owen-Smith, S. Wordsworth, J.M. Blazeby, 'Why the NHS should do more bariatric surgery; how much should we do?', *BMJ*, 2016, 353:i1472, https://www.bmj.com/content/353/bmj.i1472

14. R.C. Troke, T.M. Tan and S.R. Bloom, 'The future role of gut hormones in the treatment of obesity', *Therapeutic Advances in Chronic Disease*, January 2014,

5(1), pp. 4–14, https://www.ncbi.nlm.nih.gov/pmc/articles/PMC3871274/#b ibr61-2040622313506730

18 NON-DIETING

1. T.L. Tylka, R.A. Annunziato, D. Burgard, S. Daníelsdóttir, E. Shuman, C. Davis and R.M. Calogero, 'The Weight-Inclusive versus Weight-Normative Approach to Health: Evaluating the Evidence for Prioritizing Well-Being over Weight Loss', *Journal of Obesity*, Volume 2014, Article ID 983495, https://www.researchgate. net/publication/264987677_The_Weight-Inclusive_versus_Weight-Normative_ Approach_to_Health_Evaluating_the_Evidence_for_Prioritizing_Well-Being_ over_Weight_Loss

2. Ibid; D. Neumark-Sztainer, M. Wall, J. Haines, M. Story, M.E. Eisenberg, 'Why does dieting predict weight gain in adolescents? Findings from project EAT-II: a 5-year longitudinal study', *Journal of the American Dietetic Association*, March 2007, 107(3), pp. 448–55, https://www.ncbi.nlm.nih.gov/pubmed/17324664

3. M.L. Hatzenbuehler, J.C. Phelan and B.G. Link, 'Stigma as a Fundamental Cause of Population Health Inequalities', *American Journal of Public Health*, May 2013, 103(5), pp. 813–21, https://www.ncbi.nlm.nih.gov/pmc/articles/ PMC3682466/

4. M. Wirth, C.E. Blake, J.R. Hébert, X. Sui and S.N. Blair, 'Chronic weight dissatisfaction predicts type 2 diabetes risk: Aerobic Center Longitudinal Study', *Health Psychology*https://www.ncbi.nlm.nih.gov/pmc/articles/PMC4115022/

5. D. Vlahov and B. Junge, 'The role of needle exchange programs in HIV prevention', *Public Health Reports*, June 1998, 113(Suppl 1), pp. 75–80, https://www. ncbi.nlm.nih.gov/pmc/articles/PMC1307729/

6. E. Tribole, 'Intuitive Eating: Research Update', *SCAN's Pulse*, 2017, Volume 36, No. 3, http://www.laurathomasphd.co.uk/wp-content/uploads/2017/10/Tribole. IntuitiveEatingResearchUpdate.SCAN_.2017.pdf

7. Diabetes Prevention Program Research Group, '10-year follow-up of diabetes incidence and weight loss in the Diabetes Prevention Program Outcomes Study', *The Lancet*, 14 November https://www.ncbi.nlm.nih.gov/pmc/articles/ PMC3135022/

8. X. Pi-Sunyer, 'The Look AHEAD Trial: A Review and Discussion of Its Outcomes', *Current Nutrition Reports*, December https://www.ncbi.nlm.nih.gov/pmc/articles/PMC4339027/

9. E. M. Matheson, King and Everett, 'Healthy lifestyle habits and mortality in overweight and obese individuals', *Journal of the American Board of Family Medicine*, January–February 2012, 25(1), pp. 9-15, https://www.ncbi.nlm.nih. gov/pubmed/22218619

10. Wei et al., 'Relationship between low cardiorespiratory fitness and mortality in normal-weight, overweight, and obese men', *JAMA*, 27 October 1999, 282(16), pp. 1547–53, https://www.ncbi.nlm.nih.gov/pubmed/10546694

11. R. Puhl, Peterson and J. Luedicke, 'Fighting obesity or obese persons? Public perceptions of obesity-related health messages', *International Journal of Obesity* (London), June 2013, 37(6), pp. 774–82, https://www.ncbi.nlm.nih.gov/ pubmed/22964792

19 HOW THE FRENCH LUNCH

1. B.P. Chapman, K. Fiscella, P. Duberstein, I. Kawachi and M. Coletta, 'Can the influence of childhood socioeconomic status on men's and women's body mass be explained by socioeconomic status or personality? Findings from a national sample', *Health Psychology*2009, 28(4), pp. 419–27, https://www.ncbi.nlm.nih.gov/pubmed/19594266
2. Ibid.
3. P.L. Hill, N.A. Turiano, M.D. Hurd, D.K. Mroczek and B.W. Roberts, 'Conscientiousness and longevity: an examination of possible mediators', *Health Psychology*,2011, 30(5), pp. 536–41, https://www.ncbi.nlm.nih.gov/pubmed/21604882
4. Provencher et al., 'Personality traits in overweight and obese women: associations with BMI and eating behaviors', *Eating Behavior*,2008, 9(3), pp. 294–302, https://www.ncbi.nlm.nih.gov/pubmed/18549988/
5. B. Simini, 'Serge Renaud: from French paradox to Cretan miracle', *The Lancet*, 1 January 2000, Volume 355, No. 9197, p. 48, https://www.thelancet.com/journals/lancet/article/PIIS0140-6736(05)71990-5/fulltext
6. J. Holt-Lunstad, T.B. Smith and J.B. Layton, 'Social Relationships and Mortality Risk: A Meta-analytic Review', *PLOS Medicine*, July 2010, 7(7), e1000316, https://www.ncbi.nlm.nih.gov/pmc/articles/PMC2910600/
7. Kwon et al., 'Eating alone and metabolic syndrome: A population-based Korean National Health and Nutrition Examination Survey 2013–2014', *Obesity Research & Clinical Practice*, March–April 2018, Volume 12, Issue 2,pp. 146–57, https://www.sciencedirect.com/science/article/pii/S1871403X17300960#!

20 GOING DUTCH

1. Adab et al., 'Effectiveness of a childhood obesity prevention programme delivered through schools, targeting 6 and 7 year olds: cluster randomised controlled trial (WAVES study)', *BMJ*,2018, 360:k211, https://www.bmj.com/content/360/bmj.k211
2. P.L. Mabry, D.H. Olster, G.D. Morgan and D.B. Abrams, 'Interdisciplinarity and Systems Science to Improve Population Health – A View from the NIH Office of Behavioral and Social Sciences Research', *American Journal of Preventative Medicine*, August https://www.ncbi.nlm.nih.gov/pmc/articles/PMC2587290/
3. Government Office for Science, Foresight Report, 'Tackling Obesities: Future Choices', 2007, https://assets.publishing.service.gov.uk/government/uploads/system/uploads/attachment_data/file/287943/07-1469x-tackling-obesities-future-choices-summary.pdf
4. Z. Schlanger, 'Bloomberg's Jumbo Soda Ban for NYC is Now Officially Dead', *Newsweek*, 26 June 2014, http://www.newsweek.com/bloombergs-jumbo-soda-ban-dies-new-york-judge-rules-it-exceeds-citys-256376
5. *Off the Scales: Tackling England's Childhood Obesity Crisis*, Report from The Centre for Social Justice, 2017, https://www.centreforsocialjustice.org.uk/core/wp-content/uploads/2017/12/CSJ-Off-The-Scales-Obesity-Report.pdf

INDEX